Between Reason and Experience

Inside Technology

edited by Wiebe E. Bijker, W. Bernard Carlson, and Trevor Pinch

Janet Abbate, *Inventing the Internet*

Atsushi Akera, *Calculating a Natural World: Scientists, Engineers and Computers during the Rise of U.S. Cold War Research*

Charles Bazerman, *The Languages of Edison's Light*

Marc Berg, *Rationalizing Medical Work: Decision-Support Techniques and Medical Practices*

Wiebe E. Bijker, *Of Bicycles, Bakelites, and Bulbs: Toward a Theory of Sociotechnical Change*

Wiebe E. Bijker and John Law, editors, *Shaping Technology/Building Society: Studies in Sociotechnical Change*

Wiebe E. Bijker, Roland Bal, and Ruud Hendriks, *The Paradox of Scientific Authority: The Role of Scientific Advice in Democracies*

Karin Bijsterveld, *Mechanical Sound: Technology, Culture, and Public Problems of Noise in the Twentieth Century*

Stuart S. Blume, *Insight and Industry: On the Dynamics of Technological Change in Medicine*

Pablo J. Boczkowski, *Digitizing the News: Innovation in Online Newspapers*

Geoffrey C. Bowker, *Memory Practices in the Sciences*

Geoffrey C. Bowker, *Science on the Run: Information Management and Industrial Geophysics at Schlumberger, 1920–1940*

Geoffrey C. Bowker and Susan Leigh Star, *Sorting Things Out: Classification and Its Consequences*

Louis L. Bucciarelli, *Designing Engineers*

Michel Callon, Pierre Lascoumes, and Yannick Barthe, *Acting in an Uncertain World: An Essay on Technical Democracy*

H. M. Collins, *Artificial Experts: Social Knowledge and Intelligent Machines*

Park Doing, *Velvet Revolution at the Synchrotron: Biology, Physics, and Change in Science*

Paul N. Edwards, *The Closed World: Computers and the Politics of Discourse in Cold War America*

Andrew Feenberg, *Between Reason and Experience: Essays in Technology and Modernity*

Herbert Gottweis, *Governing Molecules: The Discursive Politics of Genetic Engineering in Europe and the United States*

Joshua M. Greenberg, *From Betamax to Blockbuster: Video Stores and the Invention of Movies on Video*

Kristen Haring, *Ham Radio's Technical Culture*

Gabrielle Hecht, *The Radiance of France: Nuclear Power and National Identity after World War II*

Gabrielle Hecht, *The Radiance of France: Nuclear Power and National Identity after World War II, New Edition*

Kathryn Henderson, *On Line and On Paper: Visual Representations, Visual Culture, and Computer Graphics in Design Engineering*

Christopher R. Henke, *Cultivating Science, Harvesting Power: Science and Industrial Agriculture in California*

Christine Hine, *Systematics as Cyberscience: Computers, Change, and Continuity in Science*

Anique Hommels, *Unbuilding Cities: Obduracy in Urban Sociotechnical Change*

Deborah G. Johnson and Jameson W. Wetmore, editors, *Technology and Society: Building Our Sociotechnical Future*

David Kaiser, editor, *Pedagogy and the Practice of Science: Historical and Contemporary Perspectives*

Peter Keating and Alberto Cambrosio, *Biomedical Platforms: Reproducing the Normal and the Pathological in Late-Twentieth-Century Medicine*

Eda Kranakis, *Constructing a Bridge: An Exploration of Engineering Culture, Design, and Research in Nineteenth-Century France and America*

Christophe Lécuyer, *Making Silicon Valley: Innovation and the Growth of High Tech, 1930–1970*

Pamela E. Mack, *Viewing the Earth: The Social Construction of the Landsat Satellite System*

Donald MacKenzie, *Inventing Accuracy: A Historical Sociology of Nuclear Missile Guidance*

Donald MacKenzie, *Knowing Machines: Essays on Technical Change*

Donald MacKenzie, *Mechanizing Proof: Computing, Risk, and Trust*

Donald MacKenzie, *An Engine, Not a Camera: How Financial Models Shape Markets*

Maggie Mort, *Building the Trident Network: A Study of the Enrollment of People, Knowledge, and Machines*

Peter D. Norton, *Fighting Traffic: The Dawn of the Motor Age in the American City*

Helga Nowotny, *Insatiable Curiosity: Innovation in a Fragile Future*

Ruth Oldenziel and Karin Zachmann, editors, *Cold War Kitchen: Americanization, Technology, and European Users*

Nelly Oudshoorn and Trevor Pinch, editors, *How Users Matter: The Co-Construction of Users and Technology*

Shobita Parthasarathy, *Building Genetic Medicine: Breast Cancer, Technology, and the Comparative Politics of Health Care*

Trevor Pinch and Richard Swedberg, editors, *Living in a Material World: Economic Sociology Meets Science and Technology Studies*

Paul Rosen, *Framing Production: Technology, Culture, and Change in the British Bicycle Industry*

Richard Rottenburg, *Far-Fetched Facts: A Parable of Development Aid*

Susanne K. Schmidt and Raymund Werle, *Coordinating Technology: Studies in the International Standardization of Telecommunications*

Wesley Shrum, Joel Genuth, and Ivan Chompalov, *Structures of Scientific Collaboration*

Charis Thompson, *Making Parents: The Ontological Choreography of Reproductive Technology*

Dominique Vinck, editor, *Everyday Engineering: An Ethnography of Design and Innovation*

Between Reason and Experience

Essays in Technology and Modernity

Andrew Feenberg

Foreword by Brian Wynne
Afterword by Michel Callon

The MIT Press
Cambridge, Massachusetts
London, England

This book was set in Stone Sans and Stone Serif by Westchester Book Group.

Library of Congress Cataloging-in-Publication Data

Feenberg, Andrew.
Between reason and experience : essays in technology and modernity / Andrew Feenberg ; foreword by Brian Wynne ; afterword by Michel Callon.
    p. cm.— (Inside technology)
Includes bibliographical references and index.
ISBN 978-0-262-51425-5 (pbk. : alk. paper) 1. Technology—Philosophy.
2. Technology—Social aspects. 3. Civilization, Modern. I. Title.
T14.F429   2010
601—dc22

                                        2009037833

Contents

Foreword

Brian Wynne

Andrew Feenberg's work exploring the various conundrums and openings in the inter-relations of science, technology, and democracy, has been one of the few bodies of thoroughgoing philosophical work that has consistently engaged in constructive struggle with sociology of scientific knowledge and technology, or what is more well-known as science and technology studies (STS). Indeed using a constructivist approach that STS has pioneered often in the teeth of mainstream philosophical complaint, Feenberg has produced many original insights showing how endemic ambiguities, incompleteness, and differences of meaning or purpose in social constructions of technology, all of which call for negotiation and flexibility, and which are in short, *political*, are routinely reduced to and enacted as matters of expert discovery and "fact." However his main contribution has not just been critical in this sense. He has patiently built a considerable body of philosophically informed work that identifies the foundations of an authentic democratic politics of technology—or as STS prefers it, of *technoscience*—as an increasingly urgent replacement of the manifestly bankrupt "unpolitics," that technoscience is claimed to play.

For these reason alone, I treat it as an honor to have been asked to write a foreword to this book of essays. In laying down some enlightened challenges to STS, Feenberg also joins with it in demolishing some of the most sacrosanct edifices of modern global capitalism's pervasive infiltration of science, rationality, and innovation. This infiltration has involved the mutual construction of science and politics in "given" but silently selective trajectories of technological innovation—and of correspondingly reductionist regulation and risk assessment—all promoted in the name of science.

In Feenberg's own words in his introduction to chapter 9:

> In modern times the new mechanistic concept of nature shattered the harmony between experience and scientific rationality. . . . The world split into two incommensurable spheres: a rational but meaningless nature and a human environment still rich in meaning but without rational foundation. In the centuries since the scientific revolution, no persuasive way has been found to validate experience or to reunite the worlds despite the repeated attempts of philosophers from Hegel to Heidegger. This is not just a theoretical problem. . . . Once the lessons of experience no longer shape technical advance, it is guided exclusively by the pursuit of wealth and power. The outcome calls into question the viability of modernity.

Thus common democratic life-world experience needs to be reconnected with the differentiated worlds of abstracted and *interested* instrumental expert reason (and power), so as to explore through a genuine democratic—and global—politics, the possible social-technical directions and distributions of innovation.

It goes almost without saying that complex and usually distributed but highly coordinated modern technologies, once established, lay down both material and imaginative pathways and constraints that themselves effectively delimit what may be seen as possible future developments. However as Feenberg shows—in excellent intellectual and political company here—this does not or should not be allowed to lead to the logical non sequitur, namely the long-discredited but perverse and persistent theology of technological determinism. Langdon Winner (1977) showed how this false theology inadvertently leads to the widespread and disempowering cultural idea that technology is "out of human control." In these essays (cf. chapter 2) Feenberg emphasizes, with due critical edge, how even what is a widely influential, perhaps the still-dominant "environmentalist" approach to urgent contemporary environmental challenges, has led itself into the same technological determinist cul-de-sac. It has done this by suggesting that environmental sustainability demands reversion rather than democratically imaginative and distributed social and technological innovation, against the concentration and concomitant exclusion—in knowledge, and in knowledge-ability, as well as in technology and resources—that global state-sponsored capitalism demands. Feenberg describes this still-influential conservative account of environmental realities as the "trade-off theory"—which posits that we must choose between environmentalism and industrialism.

Feenberg rightly criticizes economics for having powerfully encouraged this technological determinist falsehood, at least by default. One of the very few economists to have challenged this determinist account of technological innovation has been Brian Arthur (2009). As the UK-based Economic & Social Research Council's STEPS Centre research programme describes it (www.anewmanifesto.org), economics of innovation has always focused on how to achieve *more* innovation, and *faster*; but having declined to follow STS's lead and enter into technology or science as social (and economic) worlds in themselves, economics has black-boxed them as "mystery variables." Therefore it has never been able to ask informed questions of a more expressly and honestly normative kind, about what are the unrecognized flexibilities in the forms of technology we could live, and *which are the directions* in which technology should be developed for democratic, sustainable global societies?

These questions entirely correspond with Feenberg's agenda. However where he has been somewhat reserved and ambiguous about the extent to which the same applies to scientific knowledge, I would be more forthright, and suggest that for a considerable time society has been selectively directing not only technology but also (perhaps less directly so) scientific knowledge-inquiry and production—notwithstanding that lasting basic scientific understanding of nature has developed and accumulated alongside this more selectively applications-imagining, techno-scientific research activity.

I would thus add a further question here to Feenberg's enriching intellectual and political perspective. Even where serious environmental challenges are recognized for what they are, the pervasive technological-scientific obsession tends strongly to distort the imagination of societal responses in the direction only of (sophisticated) *technological* innovations. This often also means a selective focus on only big-technology, concentrated science-intensive responses; which often itself means production-side, as distinct from "demand-side" thoroughly social (or social-led technical) innovations. This concentrating syndrome, in science, technology, and innovation, can be argued to be an intrinsic function of modernity per se; but it can also be seen as a function of modern capitalism's requirement for concentration as a condition of surplus value extraction, in a knowledge-based economic era. Human and social innovations that might reduce turnover and processing of nature, while bringing positive environmental and cultural consequences, are increasingly excluded from dominant societal imaginaries, in favor of only concentrating technological "solutions." Smashing this deep trajectory and instigating in its place distributed and diversely grounded, pluralistic and hybrid reason-informed innovation cultures would be an alternative democratic modernity.

Feenberg argues, from a sympathetic position, that STS is an essential resource in this possible liberation; but he pleads for its adoption of a more ambitiously normative standpoint in order to make this historic contribution. His disappointment with STS is that for all its important work showing the multiple indeterminacies of technological development, and the corresponding unseen flexibilities for other innovation directions, it has shied away from modernity theory as such, including neo-Marxism. As he sees it, STS has thus risked inadvertent collapse into anti-modernity regression, or post-modernist solipsism, rather than as Feenberg wishes for, a full-frontal struggle for modernity's democratic cosmopolitan soul, including its increasingly constitutional and crucial *technological* dimensions. As with much STS work, he illuminates the historically suppressed democratic opportunities in science and technology here, pointing to the ways in which democratic social values and needs can impinge on technological imaginations, choices, and designs, by bringing reason and modern processes of social rationalization into more deliberately constructive encounters with democratic life-world experiences and meanings in all their grounded diversity. His case study (chapter 5) of the French collective experiment with Videotex and Minitel is a classic in this respect. This emphasis on the essentially arbitrary character of the technological-social constitutions into which we find ourselves "locked" by political artifice, resonates fully with STS, and with cutting-edge economic work on innovation (Arthur 2009)

In this explicitly committed philosophical exposition drawing with critical discrimination on inter alia Heidegger, Weber, Marcuse, Adorno, and Habermas, Feenberg leaves some intriguing ambiguities as to how he understands the key agent here—science. To what extent is science also subject to the same democratic considerations which Feenberg brings to technology and its directions and limits—and possibilities? Perhaps this ultimately difficult question can be answered by default—attempt the revisions and political-economic innovations with respect to technology which Feenberg advocates, and science will look after itself? It has in any case always been imbued with its own "social" and "cultural," even while operating in its differentiated specialist ways from society and politics at large. The prevailing demand for the awesome extent of proliferating science funding to provide a quid pro quo of beneficial social-economic impact, only intensifies the political economy of promise of which scientific research is the central currency. Untutored and unaccountable imaginations of future societal pay-offs, needs, and priorities, shape material scientific and technological commitments and learning trajectories, just as they preempt and starve others that could have been pursued. A democratization of such guiding imaginations, something Feenberg thus-far only alludes to, would also be an important research and collective experimentation agenda for a democratic technology politics and philosophy to pursue. Epistemic profiles in scientific research culture—for example, precision and control (thus concomitant silent externalisations) prioritized as "good science" over comprehensiveness and scope; or a focus on what is manipulable and buildable, hence exploitable within current theoretical horizons, rather than "what lies beyond" current knowledge—are technical *but also social* issues. They are properly amenable to social debate and influence; and they have social consequences.

Thus the role of science as it has come to play such a crucial defining role in capitalist idioms of "knowledge economy" and the fractured and insecure social relations so produced, has become increasingly to act as gatekeeper of democratic imaginations of possible change. Furthermore, in both its social scientific (economistic and rational choice models) and natural scientific forms, it has also become a powerful author of dominant understandings of human relations and potentialities, subordinated as these are to unduly narrow science-defined but normative policy and commercial models of "innovation," and "rationality." *How* far, and in what ways a proper democratic politics of technology and innovation would reshape what we know as science, even in its so-called pure, basic form, remains an open question. But we can say with some confidence that a deliberate democratic *(re)design* of that science would be futile; better perhaps to leave the boundaries between technology and science alone, and let science reach its own accommodations, its own normatively weighted imaginations, and its own epistemic cultures, once we have achieved more democratically mature, open and dynamic forms of technology.

A final issue, which perhaps Feenberg will illuminate in future work, pertains to this imagination of a democratized technology, not only in relation to scientific knowledge and its non-neutral normative power, but in relation to human life-worlds themselves. He talks of the systematic reductionism involved in the "primary instrumentalizations" of technical choices, codes, and designs, before these encounter the further social worlds of users and reactions, where in his terms, "secondary instrumentalizations" or redesigns, occur. He then valuably analyzes and exemplifies the opportunities for democratic reshaping which such life-world experience and values can and do bring to bear on the technologies (and their technical codes and standards) in question. This is valuable, and opens many doors both for STS scholars, and also for practical democratic initiatives. The questions that remain here, concern how a democratic and environmentally sustainable innovation may have to encompass not only new technological directions and designs influenced by more enlightened normative commitments, but also new social directions which de facto require *less* technological activity, thus less resource-concentration and inequity, and less environmental "turnover" consumption, and destruction. It appears to remain a challenge to Feenberg's democratic technology theory, as to how democratic enhancement could also potentially encompass, and maybe require, diminished aggregate technological activity, as an *increase* of social welfare. This may also be a place where Latour's attempt to reconstruct the very categories of modernity, nature, and culture, an attempt which Feenberg criticizes, have some useful purchase. The most obvious examples here would be lifestyle and social-relational changes in response to climate change and excessive greenhouse gas emissions, which would reduce energy consumption, as distinct from solely production-side technology changes or redesigns. It is not clear, in other words, how democratic technology innovation could, when appropriate, disinvent *itself*, as technology, in the interests of a better and environmentally sustainable and just democracy? Especially in a globalized world where we cannot see or feel the distant impacts of our own local actions—and where the favourite response to almost every issue we face seems to be to consume more— this becomes a more urgent kind of question.

Feenberg's important conceptual move to reemphasise meanings, as distinct from functions (to which he rightly notes, they are typically reduced by philosophers—and I would add, social scientists—as much as by technical experts), seems to offer some constructive room for response here. It is past time to wrest the authorship of public meanings from science, which has been handed this role by cumulating political default of the late twentieth century, back into democratic responsibility. Hannah Arendt (2005) recognized this default, and the risks to democracy that this exaggerated, subtly different form of dependency of modernity upon science could inflict. I look forward to future work addressing those challenges, not only by philosophers, STS analysts and others working in tandem, but also by democratic political practitioners, inside and outside our existing political institutions.

References

Arendt, Hannah. 2005. *The Promise of Politics*. New York: Schocken Books.

Arthur, W. Brian. 2009. *The Nature of Technology: What It Is and How It Evolves*. New York: Free Press.

Winner, Langdon. 1977. *Autonomous Technology*. Cambridge, MA: MIT Press.

Preface

Technical creation involves interaction between reason and experience. Knowledge of nature is required to make a working device. This is the element of technical activity we think of as rational. But the device must function in a social world, and the lessons of experience in that world influence design.

In premodern societies technical development was shaped by experience through craft traditions that combined many different registers of phenomena: religious prohibitions, practical lessons, taste, and age and gender roles. Technique was channeled into paths compatible with the local religious beliefs and customs in which the lessons of experience were conserved. Craft also combined knowledge of nature seamlessly with what the community had learned about the disruptive potential of technical achievements. Although some major failures occurred, for example, the gradual deforestation of much of the land bordering on the Mediterranean, on the whole this technical activity was compatible with stable societies that reproduced themselves more or less unchanged for generations.

The modern world develops a technology increasingly alienated from everyday experience. This is an effect of capitalism that restricts control of design to a small dominant class and its technical servants. The alienation has the advantage of opening up vast new territories for exploitation and invention, but there is a corresponding loss of wisdom in the application of technological power. The new masters of technology are not restrained by the lessons of experience and accelerate change to the point where society is in constant turmoil.

Not only is the role of experience in technical affairs reduced, but even where it still has impacts they are frequently invisible. Technology is perceived as autonomous, and technical disciplines present the effects of past social influences as purely rational specifications. Many technical standards depend on taste, but we are hardly aware of their source until we visit a country with different standards. No technical logic presides over differences in such things as domestic architecture, lighting, the normal height of tables and chairs, the placement of items on the automotive dashboard. Other standards change as health or environmental concerns are articulated and as legislation regulates industrial processes. Soon we forget the origin in public demands of the new methods and devices.

Even medical procedures evolve under the impact of experience. Consider the huge variations in obstetrics from one time and place to another. Not so long ago husbands paced back and forth in waiting rooms while their wives gave birth under anesthesia. Today husbands are invited into labor and delivery rooms, and women encouraged to rely less on anesthetics. The result of scientific discoveries? Hardly. But in both cases the system is medically prescribed and the feminist and natural childbirth movements of the 1970s that brought about the change forgotten. A technological unconscious hides the interaction between reason and experience.

This unconscious masks another important aspect of the modern institution of technology. In traditional societies social identities are stable since the social world is stable. But modern societies construct and destroy worlds and their associated identities at the rhythm of technological change. The extent of the dependency of social groups on the technological underpinnings of their world suddenly becomes visible at the moment of collapse but then quickly fades from view again. This is most obvious when changes in technology eliminate skilled crafts or restructure organizations. Worlds change with technology, and soon the orphaned identities remain alive only in the memories of the victims.

Still more obscure are the processes that generate temporary groups alarmed at new technological risks, but they are becoming more and more important to the future of technologically advanced societies. Take the exemplary case of Love Canal. The inhabitants of this upstate New York neighborhood discovered that their illnesses were caused by a new element in their world, a toxic element boiling up from the waste dump on which their houses were situated. This discovery about the world was also a self-discovery: these neighbors had suddenly become actors in a host of new relationships to scientists, the government, and the corporate author of their misfortune. Understanding of the world and identity go hand in hand. Both are fluid in modern societies, and both are intertwined with technology.

These examples illustrate the social character of technology. The idea of a pure technological rationality that would be independent of experience is essentially theological. One imagines a hypothetical infinite actor capable of a "do from nowhere."[1] God can act on his objects without reciprocity. He creates the world without suffering any recoil, side effects, or blowback. He is at the top of the ultimate practical hierarchy, in a one-way relation to his realm, not involved with things and exposed to their independent power. He has nothing like what we call "experience."

Modern philosophy takes this imaginary relation as the model of rationality and objectivity, the point at which humanity transcends itself in pure thought. But in reality we are not gods. Human beings can act only on a system to which they themselves belong. This is the practical significance of embodiment and implies participation in a world of meanings and causal powers we do not control. Finitude shows up as the reciprocity of action and reaction. Every one of our acts returns to us in some form as feedback from our objects. This is obvious in everyday communication where anger usually evokes anger, kindness evokes kindness, and so on.

The technical subject is finite too, but the reciprocity of finite action is dissipated or deferred in such a way as to create the space of a necessary illusion of transcendence. We call an action "technical" when the actor's impact on the object is out of all proportion to the return feedback affecting the actor. But this appears to be true only from a narrow view of the process. In a larger context or a longer time frame there is always plenty of feedback. This is certainly the case with causal impacts such as pollution. Identities and meanings are also at stake in technical action.

For example, we hammer in nails, transforming a stack of lumber into a table, but we are not transformed. All we experience is a little fatigue. This typical instance of technical action is narrowly framed here to highlight the apparent independence of actor from object. In the larger scheme of things, the actor is affected by his action: he becomes a carpenter or a hobbyist. His action has an impact on his identity, but that impact is not visible in the immediate technical situation where big changes occur in the wood while it seems that the man wielding the hammer is unaffected.

This example may seem trivial, but from a systems point of view there is no difference of principle between making a table and making an atom bomb. When J. Robert Oppenheimer exploded the first bomb at the Trinity test site, he suddenly recalled a passage from the Bhagavad Gita: "I have become death, shatterer of worlds." In this case the similarity between technical labor and divine action is all too clear. Technology appears to make possible a partial escape from the human condition. But Oppenheimer was soon attempting to negotiate disarmament with the Russians. He realized the shatterer could be shattered. Presumably Shiva, the god of death, does not have this problem.

Without wishing to return to traditional arrangements, we can nevertheless appreciate their wisdom, based as they were on a longer-term view of the wider context of technology than we are accustomed to today. Tradition was overthrown in modern times and society exposed to the full consequences of rapid and unrestrained technical advance, with both good and bad results. The good results were celebrated as progress, while the unintended and undesirable consequences of technology were ignored as long as it was possible to isolate and suppress the victims and their complaints. The dissipated and deferred feedback from technical activity, such unfortunate side effects as pollution and the deskilling of industrial work, were dismissed as the price of progress. The illusion of technique became the dominant ideology.

The philosopher Martin Heidegger understands this illusion as the structure of modern experience, the way in which "being" is revealed to us. While objects enter experience only insofar as they are useful in the technological system, the human subject appears as pure disincarnated rationality, methodically controlling and planning as though external to its own world. In this book I relate what Heidegger calls the "technological revealing" not to the history of being but to the consequences of persisting divisions between classes and between rulers and ruled in the many technically mediated institutions of modern societies.

These divisions culminate in a technology cut off to a considerable extent from the experience of those who live with it and use it. But as it grows more powerful and pervasive, technology has consequences for everyone that cannot be denied. In the final analysis it is impossible to insulate technology from the demands of the underlying population. Feedback from users and victims of technology eventually affects the technical codes that preside over design. Early examples emerge in the labor movement around issues of health and safety at work. Later such issues as food safety and environmental pollution signal the widening circle of affected publics. Today these interactions are becoming routine, and new groups emerge frequently as "worlds" change.

In the literature of technology studies, this is called the "co-construction" of society and technology. The examples cited here show how technology and society "co-construct" each other in ever tighter feedback loops, like the *Drawing Hands* in M. C. Escher's famous print of that name. I want to use this image to discuss the underlying structure of the technology-society relationship.

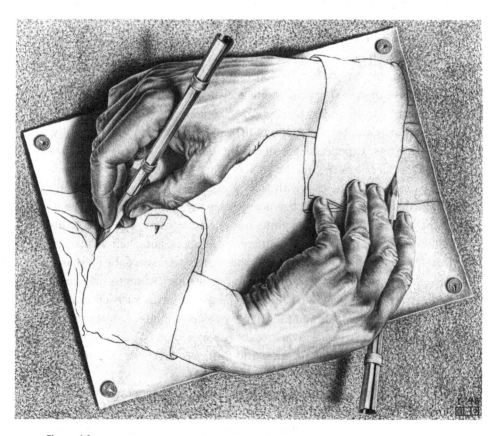

Figure I.1
M. C. Escher's *Drawing Hands*

Escher's self-drawing hands are emblematic of the concept of the "strange loop" or "entangled hierarchy" introduced by Douglas Hofstadter in his book *Gödel, Escher, Bach* (Hofstadter 1979, 10–15). The strange loop arises when moving up or down in a logical hierarchy leads paradoxically back to the starting point. Relationships between actors and their objects, such as seeing and being seen or talking and listening, are logical hierarchies in this sense. The active side stands at the top and the passive side at the bottom of these hierarchies.

In the Escher print, the paradox is illustrated in a visible form. The hierarchy of "drawing subject" and "drawn object" is "entangled" by the fact that each hand plays both functions with respect to the other (Hofstadter 1979, 689–690). If we say that the hand on the right is at the top of the hierarchy, drawing the hand on the left, we come up against the fact that the hand on the left draws the hand on the right and so is also located at the top level. Thus neither hand is at the top, or both are, which is contradictory.

As I have described it here, the relation between technical reason and experience is an entangled hierarchy. Social groups form around the technologies that mediate their relations, make possible their common identity, and shape their experience. We all belong to many such groups. Some are defined social categories, and the salience of technology to their experience is obvious. Such is the case with factory workers, whose organization and employment depend on the technology they use. Other groups are latent, unconscious of their commonalities until disaster strikes. The inhabitants of Love Canal may have been indifferent neighbors, but when toxic waste was discovered in the land they inhabited they were alerted to a shared danger. As a conscious collective, they recruited scientists to help them understand it and made demands on the government. Such encounters between the individuals and the technologies that bind them together in groups proliferate with consequences of all sorts. In every case, social identities and worlds emerge together and form the backbone of a modern society.[2]

Once formed and conscious of their identity, technologically mediated groups influence technical design through their choices and protests. This feedback from society to technology is paradoxical. Insofar as the group is constituted by the technical links that associate its members, its status is that of the "drawn" object in Escher's scheme. But it reacts back on those links in terms of its experience, "drawing" that which draws it. Neither society nor technology can be understood in isolation from each other.

Hofstadter's scheme has a limitation that does not apply in the case of technology. The strange loop is never more than a partial subsystem in a consistent, objectively conceived universe. Hofstadter evades ultimate paradox by positing an "inviolate level" of strictly hierarchical relations above the strange loop that makes it possible. He calls this level "inviolate" because it is not logically entangled with the entangled hierarchy it creates. In the case of the Escher drawing, the paradox exists only because of the unparadoxical activity of the actual printmaker Escher, who drew it in the ordinary way without himself being drawn by anyone. Escher, as Hofstadter presents him, appears as a kind of God in relation to his own artistic output, uninvolved in the contradictions of the world he creates.

But there is no equivalent of this "Escher" in the real world of co-construction, no inviolate god creating technology and society from the outside. All the creative activity takes place in a world that is itself created by that activity. Only in our fantasies do we transcend the strange loops of reason and experience. In the real world, there is no escape from the logic of finitude.

The nine chapters of this book concern various aspects of the technology/experience nexus. They introduce the main themes of critical theory of technology, the approach I have developed over the last twenty years. Critical theory of technology draws on insights from Heidegger, Foucault, the Frankfurt School, and constructivist sociology of technology. Each source contributes elements toward a better understanding of the relation between reason and experience.

This first part explores the dystopian critique of technology that arose as "progress" became identified with bureaucracy, propaganda, and genocide in the twentieth century. Scientific-technical rationality so dominates dystopia that no room is left for freedom and individuality. But this vision is fading as the paradigmatic technology of our time shifts from the industrial behemoths of the previous century to the new information technologies, especially the Internet. The Internet is not a finished product but is still in process. User initiative has played a major role in transforming its design. The environmental movement also gives rise to democratic interventions into technology. These two movements promise an end to dystopia if only we can find a way to protect and develop their liberating potential.

The second part presents methodological applications of critical theory of technology. The case of the French Minitel illustrates the social shaping of technology. An early domestic computer network, the Minitel system was subverted by hackers and transformed from an information utility into a communication medium. This part also focuses on the relationship between national culture and technical development, with Japan as an exemplary case. The discussion concerns the impact of globalization on Japanese modernization and the philosophical theories that accompanied it before World War II.

The third part treats the themes of this book at the philosophical level. Modernity and technology are indissolubly linked, but the disciplines that ought to collaborate in studying this connection have so far failed to communicate with each other. The core issue concerns the understanding of rationality as it is institutionalized in modern technologies and social systems. Understanding these peculiar modern institutions requires rethinking the connection of reason and experience. That process has already begun where it is most urgent, in relation to environmental issues. Philosophical reflection can contribute to this trend. The concluding chapter argues for informing expertise with the wisdom gained by living with technologies and their impacts. In a modern context, this cannot be accomplished by tradition but requires a more democratic technological regime. The gradual extension of democracy into the technical sphere is one of the great political transformations of our time.

The following chapters of this book are revised from previously published articles:

"Subversive Rationalization: Technology, Power, and Democracy," *Inquiry* (Sept.–Dec. 1992).

"From Information to Communication: The French Experience with Videotex," in *Contexts of Computer-Mediated Communication*, ed. M. Lea. (Harvester-Wheatsheaf, 1992).

"Looking Forward, Looking Backward: Reflections on the 20th Century," *Hitotsubashi Journal of Social Studies*, vol. 33, no. 1 (July 2001).

"Modernity Theory and Technology Studies: Reflections on Bridging the Gap," in *Modernity and Technology* (MIT Press, 2003).

"Technology in a Global World," in R. Figueroa and S. Harding, eds., *Science and Other Cultures: Issues in Philosophies of Science and Technology* (Routledge, 2003).

"Critical Theory of Technology: An Overview," *Tailor-made Bio-technologies*, vol. 1, no. 1 (April–May 2005).

"Between Reason and Experience," *Danish Philosophical Yearbook*, vol. 42 (2008).

"From the Critical Theory of Technology to the Rational Critique of Rationality," *Social Epistemology*, vol. 22, no. 1 (2008).

Acknowledgments

Many people have helped me in various ways to produce the essays collected here. I would like to thank Yoko Arisaka, Michael Benedikt, Catherine Bertho, Alison Cassells, Jean-Marie Charon, Gerald Doppelt, Arne Elias, Anne-Marie Feenberg, Simon Glynn, Marc Guillaume, Alastair Hannay, Douglas Kellner, Clive Lawson, Andrew Light, Marie Marchand, Tom Misa, Steven Moore, Robert Pippin, Hans Radder, Mayuko Uehara, and Tyler Veak.

I   Beyond Dystopia

The first chapter of this part introduces the main themes of this book: dystopia and democracy, the double aspects of technology as both technical and social, environmental reform of technical systems, and the contribution of social constructivism to philosophy of technology. The chapter argues, against technological and economic determinism, that the design of industrial society is politically contingent. In the future, those who today are subordinated to technology's rhythms and demands may be able to control it and determine its evolution. I call the process of creating such a society "democratic rationalization" because it requires technological advances imposed by wide public participation in technical decision making. The "costs" and "benefits" of such a fundamental transformation are incalculable.

The second chapter rejects a vision of environmentalism based on the notion of unavoidable trade-offs and offers a cultural approach to environmental politics. Existing technology is not the result of purely rational decisions about the most efficient way to do things but depends on social choices between alternative paths with different environmental consequences. Incorporating changed social values in future technical codes is not necessarily inefficient, as critics of environmentalism charge. Regulation can lead to technological changes that enhance economic activity rather than obstruct it, or change the understanding of the economy in ways that obviate supposed trade-offs.

The third chapter develops the discussion of the dystopian vision. The utopias and dystopias of the nineteenth and twentieth centuries imagined the fate of humanity in a society in which social relations are mediated by industrial technology. Utopian narratives depicted limits on the reach of technical systems while employing the wealth they produce to enrich leisure and support individuality. But there is no way to extend technical control without enrolling human beings in the system. The new democratic agenda is the recovery of agency in the technically mediated institutions of the society. The Internet advances this agenda because it supports interaction and participation to an unprecedented degree. The prospects for democratic rationalization of the technical system are improved by this new technology, the design of which has itself been the object of significant public interventions.

1 Democratic Rationalization: Technology, Power, and Freedom

The Limits of Democratic Theory

Technology is one of the major sources of public power in modern societies. So far as decisions affecting our daily lives are concerned, political democracy is largely overshadowed by the enormous power wielded by the masters of technical systems: corporate and military leaders and professional associations of groups such as physicians and engineers. They have far more to do with control over patterns of urban growth, the design of dwellings and transportation systems, the selection of innovations, and our experience as employees, patients, and consumers than do all the governmental institutions of our society put together.

Marx saw this situation coming in the middle of the nineteenth century. He argued that traditional democratic theory erred in treating the economy as an extrapolitical domain ruled by natural laws such as the law of supply and demand. He claimed that we will remain disenfranchised and alienated so long as we have no say in industrial decision making. Democracy must be extended from the political domain into the world of work. This is the underlying demand behind the idea of socialism.

Modern societies have been challenged by this demand for over a century. Democratic political theory offers no persuasive reason of principle to reject it. Indeed, many democratic theorists endorse it (Cunningham 1987). What is more, in a number of countries, socialist parliamentary victories or revolutions have brought to power parties dedicated to achieving it. Yet today we do not appear to be much closer to democratizing industrialism than in Marx's time.

This state of affairs is usually explained in one of the following two ways.

Technology is determining. On the one hand, common sense argues that modern technology is incompatible with workplace democracy. Democratic theory cannot reasonably press for reforms that would destroy the economic foundations of society. For evidence, consider the Soviet case: although they were socialists, Lenin and his successors did not democratize industry, and even at its most liberal, the democratization of Soviet society extended only to the factory gate. Today in the ex-Soviet Union, everyone still agrees on the need for authoritarian industrial management.

Technology is neutral. On the other hand, a minority of radical theorists claims that technology is not responsible for the concentration of industrial power. That is a political matter, due to the victory of capitalist and communist elites in struggles with the underlying population. No doubt modern technology lends itself to authoritarian administration, but in a different social context it could just as well be operated democratically.

In what follows, I will argue for a qualified version of this second position, somewhat different from both the usual Marxist and radical democratic formulations. The qualification concerns the role of technology, which I see as *neither* determining nor as neutral. I will argue that modern forms of hegemony are based on a specific type of technical mediation of a variety of social activities, whether it be production or medicine, education or the military, and that, consequently, democratization requires radical technical as well as political change.

This is a controversial position. Political theorists usually limit the proper application of the concept of democracy to the state. By contrast, I believe that unless democracy can be extended beyond its traditional bounds into the technically mediated domains of social life, its use value will continue to decline, participation will wither, and the institutions we identify with a free society will gradually disappear.

Let me turn now to the background of my argument. I will begin by presenting an overview of various theories that claim that technologically advanced societies require authoritarian hierarchy. These theories presuppose a form of technological determinism that is refuted by historical and sociological arguments I will briefly summarize. I will then present a sketch of a nondeterministic theory of modern society I call "critical theory of technology." This alternative approach emphasizes the impact of contextual aspects of technology on design ignored by the dominant view. I will argue that technology is not just the rational control of nature; both its development and impact are intrinsically social. I will then show that this view undermines the customary reliance on efficiency as an explanation of technological development in both optimistic and dystopian accounts of modernity. This conclusion, in turn, opens broad possibilities of change foreclosed by the usual understanding of technology. That argument is developed further in the following chapters.

Dystopian Modernity

Max Weber's famous theory of rationalization is the original argument against industrial democracy. The title of this chapter implies a provocative reversal of Weber's conclusions. He defined rationalization as the increasing role of calculation and control in social life, a trend leading to what he called the "iron cage" of bureaucracy (Weber 1958, 181–182). "Democratic" rationalization is thus a contradiction in terms.

Once traditionalist struggle against rationalization has been defeated, further resistance in a Weberian universe can only affirm an irrational life force against routine and drab predictability. This is not a democratic program but a romantic anti-dystopian one, the sort of thing that is already foreshadowed in Dostoyevsky's *Notes from Underground* and various back-to-nature ideologies.

My title is meant to reject the dichotomy between rational hierarchy and irrational protest implicit in Weber's position. If authoritarian social hierarchy is truly a contingent dimension of technical progress, as I believe, and not a technical necessity, then there must be an alternative rationalization of society that democratizes rather than centralizes control. We need not go underground or native to preserve threatened values such as freedom and individuality.

But the most powerful critiques of modern technological society follow directly in Weber's footsteps in rejecting this possibility. I am thinking of Heidegger's formulation of "the question of technology" and Ellul's theory of "the technical phenomenon" (Heidegger 1977; Ellul 1964). According to these theories, we have become little more than objects of technique, incorporated into the mechanism we have created. The only hope is a vaguely evoked spiritual renewal that is too abstract to inform a new technical practice.

These are interesting theories, important for their contribution to opening a space of reflection on modern technology. I will return to Heidegger's argument in the conclusion to this chapter and in the final part of this book. But first, to advance my own argument, I will concentrate on the principal flaw of dystopianism, the identification of technology in general with the specific technologies that have developed in the last two centuries in the West. These are technologies of conquest that pretend to an unprecedented autonomy; their social sources and impacts are hidden. I will argue that this type of technology is a particular feature of our society and not a universal dimension of modernity as such.

Technological Determinism

Determinism rests on the assumption that technologies have an autonomous functional logic that can be explained without reference to society. Technology is presumably social only through the purpose it serves, and purposes are in the mind of the beholder. Technology would thus resemble science and mathematics by its intrinsic independence of the social world.

Yet unlike science and mathematics, technology has immediate and powerful social impacts. It would seem that society's fate is at least partially dependent on a nonsocial factor that influences it without suffering a reciprocal impact. This is what is meant by "technological determinism." A deterministic view of technology is commonplace in business and government, where it is often assumed that technical progress is an exogenous force influencing society rather than an expression of changes in culture and values.

Dystopian visions of modernity are also deterministic. If we want to affirm the democratic potentialities of modern industrialism, we will therefore have to challenge their deterministic premises, the thesis of unilinear progress, and the thesis of determination by the base.

1. According to determinism, technical progress follows a unilinear course, a fixed track, from less to more advanced configurations. Although this seems obvious from a backward glance at the development of any familiar technical object, in fact it is based on two claims of unequal plausibility: first, that technical progress proceeds from lower to higher levels of

development; and second, that that development follows a single sequence of necessary stages. As we will see, the first claim is independent of the second and not necessarily deterministic.

2. Determinism also affirms that social institutions must adapt to the "imperatives" of the technological base. This view, which no doubt has its source in a certain reading of Marx, is now part of the common sense of the social sciences (Miller 1984, 188–195). Following and in the next chapter, I will discuss one of its implications in detail: the supposed "trade-off" between prosperity and environmental values.

These two theses of technological determinism present decontextualized, self-generating technology as the foundation of modern society. Determinism thus implies that our technology and its corresponding institutional structures are universal, indeed, planetary in scope. There may be many forms of tribal society, many feudalisms, even many forms of early capitalism, but there is only one modernity, and it is exemplified in our society for good or ill. Developing societies should take note: as Marx once said, calling the attention of his backward German compatriots to British advances: *"De te fabula narratur"*—of you the tale is told (Marx 1906 reprint, 13).

Constructivism

The implications of determinism appear so obvious that it is surprising to discover that neither of its two theses withstands close scrutiny. Yet contemporary sociology undermines the thesis of unilinear progress, while historical precedents are unkind to the thesis of determination by the base.

Recent constructivist sociology of technology grows out of social studies of science (Bloor 1991, 175–179; Latour 1987). I employ the term "constructivism" loosely to refer to the theory of large-scale technical systems, social constructivism, and actor-network theory. They have in common an emphasis on the social contingency of technical development. They challenge the traditional view of the autonomy of technology and study it much as one might an institution or a law. The specifics of these methodologies are not relevant here, but this general approach lends support to the critical theory of technology.

Constructivism challenges our tendency to exempt scientific theories from the sort of sociological examination to which we submit nonscientific beliefs. It affirms the "principle of symmetry," according to which all contending beliefs are subject to the same type of social explanation regardless of their truth or falsity. A similar approach to technology rejects the usual assumption that technologies succeed on purely functional grounds.

Constructivism argues that theories and technologies are underdetermined by scientific and technical criteria. Concretely, this means two things: first, there is generally a surplus of workable solutions to any given problem, with social actors making the final choice among several viable options; and second, the problem definition often changes in the course of solution.

Trevor Pinch and Wiebe Bijker illustrate these points with the example of the bicycle. In the late nineteenth century, before the present form of the bicycle was fixed, design was pulled in several different directions. Some customers perceived bicycling as a competitive sport, while others had an essentially utilitarian interest in transportation. Designs corresponding to the first definition had high front wheels that were rejected as unsafe by the second type of rider. They preferred the "safety" with two equal-sized low wheels. With the introduction of inflatable tires the low wheelers won out, and the entire later history of the bicycle down to the present day stems from that line of technical development. Technology is not determining in this example; on the contrary, the "different interpretations by social groups of the content of artifacts lead via different chains of problems and solutions to different further developments" (Pinch and Bijker 1989, 42).

Pinch and Bijker call this variability of goals the "interpretative flexibility" of technologies. What a technology *is* depends on what it is *for*, and that is often in dispute. The flexibility of technologies is greatest at the outset and diminishes as the competition between alternatives is sorted out. Finally, closure is achieved in the consolidation of a standard design capable of prevailing for an extended period. This is what happened to the bicycle, the automobile, and most of the familiar technologies that surround us.

In the case of the bicycle, the "safety" design won out, and it benefited from all the later advances. In retrospect, it seems as though the high

wheelers were a clumsy and less efficient stage in a progressive development leading through the old "safety" bicycle to current designs. In fact
the high wheeler and the "safety" shared the field for years, and neither
was a stage in the other's development. The high wheeler represents a possible alternative path of bicycle development that addressed different
problems at the origin. The defeated alternative was left frozen in time like
a dinosaur fossil and so appears obviously inferior today in a typical illusion of progress.

Determinism is a species of Whig history that tells the story as though
the end was inevitable by projecting the abstract technical logic of the finished object back into the past as the *telos* of development. That approach
confuses our understanding of the past and stifles the imagination of a different future. Constructivism can open up that future, although its practitioners have hesitated so far to engage the larger social issues implied in
their method.[1]

Indeterminism

If the thesis of unilinear progress falls, the collapse of the notion of determination by the technological base cannot be far behind. Yet it is still
frequently invoked in contemporary political debates.

I shall return to these debates later in this chapter. For now, let us consider the remarkable anticipation of current attitudes in the struggle over
the length of the workday and child labor in mid-nineteenth-century England. The debate on the Factory Bill of 1844 was entirely structured around
the opposition of technological imperatives and ideology. Lord Ashley, the
chief advocate of regulation, protested that "The tendency of the various
improvements in machinery is to supersede the employment of adult males,
and substitute in its place, the labour of children and females. What will
be the effect on future generations, if their tender frames be subjected,
without limitation or control, to such destructive agencies?"[2]

He went on to deplore the decline of the family consequent upon the
employment of women, which "disturbs the order of nature" and deprives
children of proper upbringing. "It matters not whether it be prince or
peasant, all that is best, all that is lasting in the character of a man, he has
learnt at his mother's knees." Lord Ashley was outraged to find that
"females not only perform the labour, but occupy the places of men; they

are forming various clubs and associations, and gradually acquiring all those privileges which are held to be the proper portion of the male sex. . . . they meet together to drink, sing, and smoke; they use, it is stated, the lowest, most brutal, and most disgusting language imaginable . . ."

Proposals to abolish child labor met with consternation on the part of factory owners, who regarded the little worker as an "imperative" of the technologies created to employ him. They denounced the "inefficiency" of using full-grown workers to accomplish tasks done as well or better by children, and they predicted all the usual catastrophic economic consequences—increased poverty, unemployment, loss of international competitiveness—from the substitution of more costly adult labor. Their eloquent representative, Sir J. Graham, therefore urged caution: "We have arrived at a state of society when without commerce and manufactures this great community cannot be maintained. Let us, as far as we can, mitigate the evils arising out of this highly artificial state of society; but let us take care to adopt no step that may be fatal to commerce and manufactures."

He further explained that a reduction in the workday for women and children would conflict with the depreciation cycle of machinery and lead to lower wages and trade problems. He concluded that "in the close race of competition which our manufacturers are now running with foreign competitors . . . such a step would be fatal. . . ." Regulation, he and his fellows maintained in words that echo still, is based on a "false principle of humanity, which in the end is certain to defeat itself." One might almost believe that Ludd had risen again in the person of Lord Ashley: the issue is not really the regulation of work, "but it is in principle an argument to get rid of the whole system of factory labour." Similar protestations are heard today on behalf of industries threatened with what they call environmental "Luddism."

Yet what actually happened once the regulators imposed limitations on the workday and expelled children from the factory? Did the violated imperatives of technology come back to haunt them? Not at all. Regulation led to an intensification of factory labor that was incompatible with the earlier conditions in any case. Children ceased to be workers and were redefined socially as learners and consumers. Consequently, they entered the labor market with higher levels of skill and discipline that were soon presupposed by technological design. A vast historical process unfolded, partly stimulated by the ideological debate over how children should be raised and partly economic. It led eventually to the current situation in which nobody dreams of returning to cheap child labor in order to cut costs, at least not in the developed countries.

This example shows the tremendous flexibility of the technical system. It is not rigidly constraining but on the contrary can adapt to a variety of social demands. This conclusion should not be surprising, given the responsiveness of technology to social redefinition discussed previously. In sum, technology is just another dependent social variable, albeit an increasingly important one, and not the key to the riddle of history.

Determinism, I have argued, is characterized by the principles of unilinear progress and determination by the base; if determinism is wrong, then research must be guided by two contrary principles. In the first place, technological development is not unilinear but branches in many directions and could reach generally higher levels along several different tracks. In the second place, technological development is not determining for society but is overdetermined by both technical and social factors.

The political significance of this position should also be clear by now. In a society where determinism stands guard on the frontiers of democracy, indeterminism "enlarges the field of the possible."[3] If technology has many unexplored potentialities, no technological imperatives dictate the current social hierarchy. Rather, technology is a scene of social struggle, a "parliament of things," on which civilizational alternatives contend (Latour 1993).

Interpreting Technology

In the next sections of this chapter, I would like to present several major themes of a nondeterminist approach to technology. The picture sketched so far implies a significant change in definition. Technology can no longer be considered as a collection of devices nor, more generally, as the sum of rational means. These are tendentious definitions that beg the question of technology's social significance and involvements.

Insofar as it is social, technology ought to be subject to interpretation like any other cultural artifact, but it is generally excluded from humanistic study. We are assured that its essence lies in a technically explainable function rather than a hermeneutically interpretable meaning. At most, humanistic methods might illuminate extrinsic aspects of technology, such as packaging and advertising, or popular reactions to controversial innovations such as nuclear power. Technological determinism draws its force from this attitude. If one ignores most of the connections between technology and society, it is no wonder that technology then appears to be self-generating.

Technical objects have two hermeneutic dimensions that I call their *social meaning* and their *cultural horizon*.[4] The role of social meaning is clear in the bicycle case. We have seen that the design of the bicycle was decided by a contest of interpretations: Was it to be a sportsman's toy or a means of transportation? Design features such as wheel size signified it as one or another type of object while also suiting it to its function.

It might be objected that this is merely an initial disagreement over functions with no hermeneutic significance. Once the object is stabilized, the engineer has the last word on its nature, and the humanist interpreter is out of luck. This is the view of most engineers and managers; they readily grasp the concept of "function," but they have no use for "meaning."

In fact the dichotomy of function and meaning is a product of modern technical cultures, which are themselves rooted in the structure of the modern economy. The concept of "function" strips technology bare of social contexts, focusing engineers and managers on just what they need to know to do their job. A fuller picture is conveyed, however, by studying the social role of technical objects and the lifestyles they make possible. That picture places the abstract notion of "function" in its concrete social context. It makes technology's contextual causes and consequences visible rather than obscuring them behind an impoverished functionalism.[5]

The functionalist point of view yields a decontextualized temporal cross-section in the life of the object. As we have seen, determinism claims implausibly to be able to get from one such momentary configuration of the object to the next on purely technical terms. But in the real world all sorts of unpredictable attitudes crystallize around technical objects and influence later design changes. The engineer may think these are extrinsic to the device he or she is working on, but they are its very substance as a historically evolving phenomenon.

These facts are recognized to a certain extent in the technical fields themselves. With computers, we have a contemporary version of the dilemma of the bicycle discussed earlier. Progress of a generalized sort in speed, power, and memory goes on apace while corporate planners struggle with the question of what it is all for. Technical development does not point definitively toward any particular path. Instead, it opens branches, and the final determination of the "right" branch is not within the competence of engineering because it is simply not inscribed in the nature of the technology.

I have studied a particularly clear example of the complexity of the relation between the technical function and meaning of the computer in the case of French videotex.[6] Called "Teletel," this system was designed to bring France into the Information Age by giving telephone subscribers access to databases through a standard dumb terminal. Fearing that consumers would reject anything resembling office equipment, the telephone company attempted to redefine the computer's social image; it was no longer to appear as a filing and calculating device for professionals but was to become a public informational network.

The telephone company designed a new type of terminal, the Minitel, to look and feel like an adjunct to the domestic telephone. The telephonic disguise suggested to some users that they ought to be able to talk to each other on the network. Soon the Minitel underwent a further redefinition at the hands of these users, many of whom employed it primarily for anonymous online chatting with other users in the search for amusement, companionship, and sex.

Thus the design of the Minitel invited communications applications that the company's engineers had not intended when they set about improving the flow of information in French society. Those applications, in turn, connoted the Minitel as a means of personal encounter, the very opposite of the rationalistic project for which it was originally created. The "cold" computer became a "hot" new medium.

At issue in the transformation was not only the computer's narrowly conceived technical function but also the very nature of society it makes possible. Does networking open the doors to the Information Age, where, as rational consumers hungry for data, we pursue strategies of optimization? Or is it a postmodern technology that emerges from the breakdown of institutional and sentimental stability? In this case technology is not merely the servant of some predefined social purpose; it is an environment within which a way of life is elaborated.

In sum, differences in the way social groups interpret and use technical objects are not merely extrinsic but make a difference in the nature of the objects themselves. *What* the object *is* for the groups that ultimately decide its fate determines what it *becomes* as it is redesigned and improved over time. If this is true, then we can understand technological development only by studying its meaning for the various groups that influence it.

Technological Hegemony

In addition to the assumptions about individual technical objects we have been discussing so far, meanings belonging to the cultural horizon of society also shape technologies. This second hermeneutic dimension of technology is the basis of modern forms of social hegemony; it is particularly relevant to our original question concerning the inevitability of hierarchy in technological society.

As I will use the term, hegemony is a form of domination so deeply rooted in social life that it seems natural to those it dominates. One might also define it as that aspect of the distribution of social power that has the force of culture behind it.

The term "horizon" refers to culturally general assumptions that form the unquestioned background to every aspect of life.[7] Some of these support the prevailing hegemony. For example, in feudal societies, the "chain of being" established hierarchy in the fabric of God's universe and protected the caste relations of the society from challenge. Under this horizon, peasants revolted in the name of the king, the only imaginable source of power. Rationalization is our modern horizon, and technological design is the key to its effectiveness as the basis of modern hegemonies.

Technological development is constrained by cultural norms originating in economics, ideology, religion, and tradition. I discussed earlier how assumptions about the age composition of the labor force entered into the design of nineteenth-century production technology. Such assumptions seem so natural and obvious they often lie below the threshold of conscious awareness.

This is the point of Herbert Marcuse's important critique of Max Weber's theory of rationalization (Marcuse 1968). Marcuse shows that Weber confounds the control of labor by management with the control of nature by technology. The search for control of nature is generic, but management arises only against a specific social background, the capitalist system. Workers have no immediate interest in output in this system since their wage is not essentially linked to the income of the firm. Control of human beings becomes all-important in this context. Another way to put it would be to say that top down management is "rational" under the horizon of capitalism, but Weber left off the qualifying phrase.

Through mechanization, some of the control functions are eventually transferred from human overseers and parcelized work practices to machines. Machine design is thus socially relative in a way that Weber never recognized, and the "technological rationality" it embodies is not universal but particular to capitalism. In fact, it is the horizon of all the existing industrial societies, communist as well as capitalist, insofar as they are managed from above.

If Marcuse is right, it ought to be possible to trace the impress of class relations in the very design of production technology, as has indeed been shown by such Marxist students of the labor process as Harry Braverman and David Noble (Braverman 1974; Noble 1984). The assembly line offers a particularly clear instance because its design achieves traditional management goals, such as deskilling and pacing work. Its technologically enforced labor discipline increases productivity and profits by increasing control. However, the assembly line appears as technical progress only in a specific social context. It would not be perceived as an advance in an economy based on workers' cooperatives in which labor discipline was largely self-imposed rather than imposed from above. In such a society, a different technological rationality would dictate different ways of increasing productivity.

This example shows that technological rationality is not merely a belief, an ideology, but is effectively incorporated into the structure of machines. Machine design mirrors back the social factors operative in the prevailing rationality. The fact that the argument for the social relativity of modern technology originated in a Marxist context has obscured its most radical implications. We are not dealing here with a mere critique of the property system but have extended the critique down into the technical "base," the forces of production. This approach goes well beyond the old economic distinction between capitalism and socialism, market and plan. Instead, one arrives at a very different distinction between societies in which power rests on the technical mediation of social activities and those that democratize technical control and, correspondingly, technological design.

Double Aspect Theory

The argument to this point might be summarized as a claim that social meaning and functional rationality are inextricably intertwined dimensions of technology. They are not ontologically distinct with meaning in the observer's mind and rationality in the technology proper. Rather they are "double aspects" of the same underlying technical object, each aspect revealed by a specific contextualization.[8]

Functional rationality isolates objects from their original context in order to incorporate them into a theoretical system. The institutions that support this procedure, such as laboratories and research and design centers, themselves form a special context with their own practices and links to various social agencies and powers. The notion of "pure" rationality arises when the work of decontextualization is not itself grasped as a social activity reflecting social interests.

Technologies are selected by these interests from among many possible configurations. Guiding the selection process are social codes established by the cultural and political struggles that define the cultural horizon under which the technology will fall. Once introduced, technology offers a material validation of the social order to which it has been preformed. I call this the "bias" of technology: apparently neutral, functional rationality is enlisted in support of a hegemony. The more technology society employs, the more significant is this support.[9]

As Foucault argued in his theory of "power\knowledge," modern forms of oppression are based not so much on false ideologies as on the specific technical "truths" that found and reproduce the dominant hegemony (Foucault 1977). So long as the contingency of the choice of "truth" remains hidden, the deterministic image of a technically justified social order is projected.

The legitimating effectiveness of technology depends on unconsciousness of the cultural horizon under which it was designed. A recontextualizing critique of technology can uncover that horizon, demystify the illusion of technical necessity, and expose the relativity of the prevailing technical choices. A politics of technology can demand changes reflecting the critique.

The possibility of such a politics is rooted in a peculiar feature of the double aspects of technology. Although function and meaning are analytically distinct aspects of technologies in any temporal cross-section, they interact externally in historical time. They enjoy what might be called a "relationship of double ingression," the data of each invading the other and operating in a paradoxical way on the other's terrain. Everyday experience, the domain of social meaning, is governed by a different logic from the scientific and engineering rationality that presides over the functional logic of technology. Where these contexts are out of alignment, tensions arise that are resolved in the course of history by changes and adjustments in one or both of them.

This is methodologically puzzling but obvious in specific cases. For example, knowledge of risk enters experience as fear or anxiety, that is, an aspect of the meaning of the associated objects. Nuclear power is a case in point. The social meaning of the technology is informed in part by scientific knowledge of risk. But more ancient layers of meaning crystallize around invisible threats and fear of the unknown. Meanwhile, scientists and engineers respond to public perceptions of risk with new designs that promise improved safety. Thus the social meaning of the technology influences the rational specification of the device. In other fields such as computing, new functionalities are routinely introduced in response to changes in meaning.

The Social Relativity of Efficiency

These issues appear with particular force in the environmental movement. Many environmentalists argue for technical changes that would protect nature and in the process improve human life as well. Such changes would enhance efficiency in broad terms by reducing harmful and costly side effects of technology. However, this program is very difficult to implement in a capitalist society. There is a tendency to deflect criticism from technological processes to products and people, from apriori prevention to aposteriori clean-up. These preferred strategies are costly and reduce efficiency in the short run. This situation has political consequences.

Reducing side effects and restoring the environment are forms of collective consumption, financed by taxes or higher prices. These approaches dominate public awareness. This is why environmentalism is generally perceived as a cost involving trade-offs and not as a rationalization increasing overall well-being. But in a society obsessed by private consumption, that perception is damning. Economists and businesspeople are fond of explaining the price we must pay in inflation and unemployment for worshipping at Nature's shrine instead of Mammon's. Poverty awaits those who will not adjust their social and political expectations to technological imperatives.

This trade-off approach has environmentalists grasping at straws for a strategy. Some hold out the pious hope that people will turn from economic to spiritual values in the face of the mounting problems of industrial society. Others expect enlightened dictators to impose technological reform on an irrational populace. It is difficult to decide which of these solutions is more improbable, but both are incompatible with basic democratic values (Heilbroner 1975).

The trade-off approach confronts us with dilemmas—environmentally sound technology versus prosperity, workers' satisfaction and control versus productivity, and so forth—where what we need are syntheses. Unless the problems of modern industrialism can be solved in ways that both protect nature and win public support, there is little reason to hope that they will ever be solved. But how can technological reform be reconciled with prosperity when it places a variety of new limits on the economy?

The child labor case shows how apparent dilemmas arise on the boundaries of cultural change, specifically, where the social definition of major technologies is in transition. In such situations, social groups excluded from the original design process articulate their unrepresented interests politically. New values the outsiders believe would enhance their welfare appear as mere ideology to insiders who are adequately represented by the existing designs.

This is a difference of perspective, not of nature. Yet the illusion of essential conflict is renewed whenever major social changes affect technology. At first, satisfying the demands of new groups after the fact has visible costs and, if it is done clumsily, will indeed reduce efficiency until better designs are found. But usually better designs are found, and what appeared to be insuperable obstacles to growth dissolve in the face of technological change.

This situation indicates the essential difference between economic exchange and technique. Exchange is all about trade-offs: more of A means less of B. But the aim of technical advance is precisely to avoid such dilemmas with what Simondon calls "concretizations," elegant designs that optimize several variables at once. A single cleverly conceived mechanism may correspond to many different social demands, one structure to many functions. Design is not a zero-sum economic game but an ambivalent cultural process that serves a multiplicity of values and social groups without necessarily sacrificing efficiency.[10]

The Technical Code

That these conflicts over social control of technology are not new can be seen from the interesting case of the "bursting boilers" (Burke 1972). Steamboat boilers were the first technology regulated in the United States. In the early nineteenth-century the steamboat was a major form of transportation, similar to the automobile or airlines today. The United States was a big country without paved roads but with many rivers and canals, hence the reliance on steamboats. But steamboats frequently blew up when the boilers weakened with age or were pushed too hard. After several particularly murderous accidents in 1816, the city of Philadelphia consulted with experts on how to design safer boilers. This was the first time an American governmental institution interested itself in the problem. In 1837, at the request of Congress, the Franklin Institute issued a detailed report and recommendations based on rigorous study of boiler construction. Congress was tempted to impose a safe boiler code on the industry, but boilermakers and steamboat owners resisted, and the government hesitated to interfere with private property.

It took from that first inquiry in 1816 until 1852 for Congress to pass effective laws regulating the construction of boilers. In that time five thousand people died in steamboat accidents. Is this many casualties or few? Consumers evidently were not too alarmed to travel on the rivers in ever increasing numbers. Understandably, the ship owners interpreted this as a vote of confidence and protested the excessive cost of safer designs. Yet politicians also won votes by demanding safety.

The accident rate fell dramatically once thicker walls and safety valves were mandated. Legislation would hardly have been necessary to achieve this outcome had it been technically determined. But in fact boiler design was relative to a social judgment about safety. That judgment could have been made on strictly market grounds, as the shippers wished, or politically, with differing implications for technical design. In either case, those results *constitute* a proper boiler. What a boiler "is" was thus defined through a long process of political struggle culminating finally in uniform codes issued by the American Society of Mechanical Engineers.

This is an example of how technology adapts to social change. What I call the "technical code" of the object mediates the process. That code responds to the cultural horizon of the society at the level of technical design. Quite down-to-earth technical parameters such as the choice and processing of materials are *socially* specified by the code. The illusion of technical necessity arises from the fact that the code is thus literally "cast in iron," at least in the case of boilers.

The conservative antiregulatory approach is based on an illusion. It forgets that the design process always already incorporates standards of safety and environmental compatibility; similarly, all technologies support some basic level of user or worker initiative. A properly made technical object simply *must* meet these standards to be recognized as such. We do not treat conformity as an expensive add-on but regard it as an intrinsic cost. Raising the standards means altering the definition of the object, not paying a price for an alternative good or ideological value, as the trade-off approach holds.

But what of the much discussed cost/benefit ratio of design changes, such as those mandated by environmental or other similar legislation? These calculations have some application to transitional situations, before technological advances responding to new values fundamentally alter the terms of the problem. But all too often, the results depend on economists' very rough estimates of the monetary value of such things as a day of trout fishing or an asthma attack. If made without prejudice, these estimates may well help to prioritize policy alternatives. But one cannot legitimately generalize from such policy applications to a universal theory of the costs of regulation.[11]

Such fetishism of efficiency ignores our ordinary understanding of the concept, which alone is relevant to social decision making. In that everyday sense, efficiency concerns the narrow range of issues that economic actors routinely address. Unproblematic aspects of technology are not included. In theory one can decompose any technical object and account for each of its elements in terms of the goals it meets, whether it be safety, speed, reliability, and so forth, but in practice no one is interested in opening the "black box" to see what is inside.

For example, once the boiler code is established, such things as the thickness of a wall or the design of a safety valve appear as essential to the object. The cost of these features is not broken out as the specific "price" of safety and compared unfavorably with a more "efficient" but less secure version of the technology. Violating the code in order to lower costs is a crime, not a trade-off. And since all further progress takes place on the basis of the new safety standard, soon no one looks back to the good old days of cheaper, insecure designs.

Design standards are controversial only while they are in flux. Resolved conflicts over technology are quickly forgotten. Their outcomes, a welter of taken-for-granted technical and legal standards, are embodied in a stable code and form the background against which economic actors manipulate unstabilized aspects of technology in the pursuit of efficiency. The code is not varied in real-world economic calculations but treated as a fixed input.

Anticipating the stabilization of a new code, one can often ignore contemporary arguments that will soon be silenced by the emergence of a new horizon of efficiency calculations. This is what happened with boiler design and child labor; presumably, the current debates on environmentalism will have a similar history, and we will someday mock those who object to cleaner air as a "false principle of humanity" that violates technological imperatives.

Noneconomic values intersect the economy in the technical code. The examples we are dealing with illustrate this point clearly. The legal standards that regulate economic activity have a significant impact on every aspect of our lives. In the child labor case, regulation helped to widen educational opportunities with human consequences that are not merely economic in character. In the riverboat case, Americans chose high levels of security, and boiler design came to reflect that choice. Ultimately, this was no trade-off of one good for another but a noneconomic decision about the value of human life and the responsibilities of government.

Technology is thus not merely a means to an end; technical design standards define major portions of the social environment, such as urban and built spaces, workplaces, medical activities and expectations, life patterns, and so on. The economic significance of technical change often pales beside its wider human implications in framing a way of life. In such cases, regulation defines the cultural framework *of* the economy; it is not just an act *in* the economy.

Heidegger's "Essence" of Technology

The theory sketched here suggests the possibility of a general reform of technology. But dystopian critics object that the mere fact of pursuing efficiency or technical effectiveness already does inadmissible violence to human beings and nature. Universal functionalization destroys the integrity of all that is. As Heidegger argues, an "objectless" world of mere resources replaces a world of "things" treated with respect for their own sake as the gathering places of our manifold engagements with "Being."[12]

This critique gains force from the actual perils with which modern technology threatens the world today. But my suspicions are aroused by Heidegger's famous contrast between a dam on the Rhine and a Greek chalice. It would be difficult to find a more tendentious comparison. No doubt modern technology is immensely more dangerous than any other. No doubt it invalidates traditional meanings without providing an adequate substitute. And Heidegger is right to argue that means are not truly neutral, that their substantive content affects society independent of the goals they serve. But this content is not *essentially* destructive; rather its significance is a matter of design and social insertion.

However, Heidegger rejects any merely social diagnosis of the ills of technological societies and claims that the source of their problems dates back at least to Plato, that modern societies merely realize a *telos* immanent in Western metaphysics from the beginning. His originality consists in pointing out that the ambition to control being is itself a way of being and hence subordinate at some deeper level to an ontological dispensation beyond human control. But the overall effect of his critique is to condemn human agency, at least in modern times, and to confuse essential differences between types of technological development.

Heidegger and his followers distinguish between the *ontological* problem of technology, which can be addressed only by achieving what they call "a free relation" to technology, and the merely *ontic* solutions proposed by reformers who wish to change technology itself. This distinction may once have seemed more interesting than it does today. In effect, Heidegger is asking for nothing more than a change in attitude toward the selfsame technical world. But that is an idealistic solution in the bad sense and one that a generation of environmental activism decisively refutes.

Confronted with this argument, Heidegger's defenders usually point out that his critique of technology is not concerned merely with attitudes but with the way being "reveals" itself. Roughly translated out of Heidegger's language, this means that the modern world has a technological form in something like the sense in which, for example, the medieval world had a religious form. Form is no mere question of attitude but takes on a material life of its own: power plants are the gothic cathedrals of our time. But this interpretation of Heidegger's thought raises the expectation that he will offer criteria for a reform of technology. For example, his critique of the tendency of modern technology to accumulate and store up nature's powers suggests the superiority of another technology that would not challenge nature in Promethean fashion.

But Heidegger does not pursue this line. Instead, he develops his argument at such a high level of abstraction he literally cannot discriminate between electricity and atom bombs, agricultural techniques and the Holocaust. In a 1949 lecture, he asserted: "Agriculture is now the mechanized food industry, in essence the same as the manufacturing of corpses in gas chambers and extermination camps, the same as the blockade and starvation of nations, the same as the production of hydrogen bombs" (quoted in Rockmore 1992, 241). All are merely different expressions of the identical "enframing" that we are called to transcend through the recovery of a deeper relation to being. And since Heidegger rejects technological regression while leaving no room for reform, it is difficult to see in what that relation would consist beyond a mere change of attitude.

Heidegger cannot take the notion of technological reform seriously because he reifies modern technology as something separate from society, as an inherently contextless force aiming at pure power. If this is the "essence" of technology, reform would be merely extrinsic. But at this point Heidegger's position converges with the very Prometheanism he rejects. Both depend on the narrow definition of technology that, at least since Bacon and Descartes, has emphasized its destiny to control the world to the exclusion of its equally essential contextual embeddedness. This definition reflects the capitalist environment in which modern technology first developed.

The exemplary modern master of technology is the entrepreneur, single-mindedly focused on production and profit. The enterprise is a radically decontextualized platform for action, without the traditional responsibilities for persons and places that went with technical power in the past. It is the autonomy of the enterprise that makes it possible to distinguish so sharply between intended and unintended consequences, between goals and contextual effects, and to ignore the latter.

The narrow focus of modern technology meets the needs of a particular hegemony; it is not a metaphysical condition. Under that hegemony, technological design is unusually decontextualized and destructive. Not technology but that hegemony is called to account when we point out that today technical means form an increasingly threatening life environment. It is that hegemony, as it is materialized in technology, which must be challenged in the struggle for a better society.

Democratic Rationalization

For generations, faith in progress was supported by two widely held beliefs: that technical necessity dictates the path of development and that the pursuit of efficiency provides a basis for identifying that path. I have argued here that both these beliefs are false and that, furthermore, they are ideologies employed to justify restrictions on opportunities to participate in decision making in industrial society. I conclude that a reform of technological society can support a broader range of values. Democracy is one of these values.

What does it mean to democratize technology? The problem is not primarily one of legal rights but of initiative and participation. Legal forms may eventually routinize claims that are asserted informally at first, but the forms will remain hollow unless they emerge from the experience and needs of individuals resisting a technocratic hegemony.

That resistance takes many forms, from union struggles over health and safety in nuclear power plants to community struggles over toxic waste disposal to political demands for regulation of reproductive technologies. These movements alert us to the need to take technological externalities into account and demand design changes responsive to the enlarged context revealed in that accounting.

Such technological controversies have become an inescapable feature of contemporary political life, laying out the parameters for official "technology assessment" (Cambrosio and Limoges 1991; Callon et al. 2009). They prefigure the creation of a new public sphere embracing the technical background of social life and a new style of rationalization that internalizes unaccounted costs born by "nature," that is, some-thing or -body exploitable in the pursuit of profit. Here respect for nature is not antagonistic to technology but opens a new path of development.

As these controversies become commonplace, surprising new forms of resistance and new types of demands emerge. The Minitel example is a model of this new situation. In France, the computer was politicized as soon as the government supplied the general public with a highly rationalistic information system. Users "hacked" the network in which they were enrolled and altered its functioning, introducing human communication on a vast scale where only the centralized distribution of data had been planned. The Internet has also given rise to many such innovative public reactions to technology.

Individuals who are incorporated into these new technical networks have learned to resist through the net itself in order to influence the powers that control it. This is not a contest for wealth or administrative power but a struggle to subvert the technical practices, procedures, and designs structuring everyday life.

It is instructive to compare these cases with the movement of AIDS patients for better medical care. Just as a rationalistic conception of the computer tends to occlude its communicative potentialities, so in medicine caring functions have become mere side effects of treatment, which is itself understood in technical terms. Patients become objects of this technique, more or less "compliant" to management by physicians. The incorporation of thousands of incurably ill AIDS patients into this system destabilized it and exposed it to new challenges (Feenberg 1995, chap. 5; Epstein 1996).

The key issue was access to experimental treatment. Clinical research is one way in which a highly technologized medical system can care for those it cannot yet cure. But until quite recently access to medical experiments has been severely restricted by paternalistic concern for patients' welfare. AIDS patients were able to open up access because the networks of contagion in which they were caught were paralleled by social networks that were already mobilized around gay rights at the time the disease was first diagnosed.

Instead of participating in medicine individually as objects of a technical practice, they challenged it collectively and politically. They "hacked" the medical system and turned it to new purposes. Their struggle represents a counter tendency to the technocratic organization of medicine, an attempt at a recovery of its symbolic dimension and caring functions.

As in the case of the Minitel, it is not obvious how to evaluate this challenge in terms of the customary concept of politics. Nor do these subtle struggles against the growth of silence in technological societies appear significant from the standpoint of the reactionary ideologies that contend noisily with capitalist modernism today. Yet the demand for communication that these movements represent is so fundamental that it can serve as a touchstone for the adequacy of political theories of the technological age.

These resistances, like the environmental movement, challenge the horizon of rationality under which technology is currently designed. Rationalization in our society responds to a particular definition of technology as a means to profit and power. A broader understanding of technology suggests a very different notion of rationalization based on responsibility for the human and natural contexts of technical action. I call this "democratic rationalization" because it requires technological advances that can be made only in opposition to the dominant hegemony. It represents an alternative to both the ongoing celebration of technocracy triumphant and the gloomy Heideggerian counterclaim that "Only a God can save us" (Heidegger 1993a).

Is democratic rationalization in this sense socialist? There is certainly room for discussion of the connection between this new technological agenda and the old idea of socialism. I believe there is significant continuity. In socialist theory, workers' lives and dignity stood for the larger contexts that modern technology ignores. The destruction of their minds and bodies on the workplace was viewed as a contingent consequence of capitalist technical design. The implication that socialist societies might design a very different technology under a different cultural horizon was perhaps given only lip service, but at least it was formulated as a goal.

We can make a similar argument today over a wider range of contexts in a broader variety of institutional settings with considerably more urgency. I am inclined to call such a position "socialist" and to hope that in time it can replace the image of socialism projected by the failed communist experiment.

More important than this terminological question is the substantive point. Why has democracy not been extended to technically mediated domains of social life despite a century of struggles? Is it because technology is incompatible with democracy or because it has been used to suppress it? The weight of the argument supports the second conclusion. Technology can deliver more than one type of technological civilization. We have not yet exhausted its democratic potential.

2   Incommensurable Paradigms: Values and the Environment

Introduction

In this chapter I will develop the argument presented in the introduction as it relates to environmental politics. Environmental issues turn on the question of technological change. But just how flexible are the systems and designs that prevail today? Is it economically feasible to bring technology into compliance with ever-more-stringent environmental standards? This chapter addresses these questions from the standpoint of philosophy of technology. I will argue that an unexamined concept of technology shared by many environmental activists and their adversaries locks them into unmediated opposition. A different understanding of technology changes the terms of the debate.

Costs and Benefits

In the early 1970s, Paul Ehrlich claimed that environmental crisis was caused by both economic and population growth. He advocated population control and "de-development" of the advanced societies to reduce overconsumption (Ehrlich and Harriman 1971). This suggestion found support in *The Limits to Growth*, a famous study of the prospects for industrial collapse due to resource exhaustion and pollution (Meadows et al. 1971). No-growth ideology influenced many early discussions of technology and the environment.

Echoes of these early arguments reappear now as a response to climate change. The most extreme predictions show the habitable portion of the Earth and the population shrinking. Industry disappears as fossil fuels run out. Cities collapse, and the human race returns to local self-sufficient communities sustaining themselves through farming and crafts.

Climate change is real, of course, but its consequences are not easy to foresee. We can hope that political resistance to confronting its implications will give way to active engagement as crises and problems accumulate. Then all the ingenuity of the planet will be devoted to avoiding the catastrophic outcome forecast by environmental pessimists. A different form of industrial society may emerge, more frugal in some respects but perhaps enriching in new ways as well.

This more hopeful prospect implies the possibility of alternative industrial systems with different environmental impacts. In denying this possibility, the claim that we must choose between industrial society and village life is essentially deterministic. It excludes a reform of modern industrialism leading to the invention of alternative technologies compatible with the health of the environment.

The stakes in this debate go beyond economics and ecology. The individualism and freedom we value so highly depend not only on political democracy but also on the technological accomplishments that support communication and transportation and leave time for education in childhood and beyond. Modernity and technology are mutually interdependent. It is inconceivable that people living in small impoverished villages could sustain the form of life we associate with modernity. Critics who valorize craft over modern technology, the village and local bartering over the city and worldwide trade, implicitly question our identity as modern human beings.

If regression to traditional village life is the solution, can the problem be worse? This is most people's reaction to the idea of de-development. Its main effect is to bring grist to the conservative mill of those opposed to "excessive" environmental regulation. The price of reform is obviously too high if the foundation of our society must be sacrificed for environmental quality. The common view, therefore, holds that it is better to keep our present system and live with the consequences rather than surrender all the advances of modern life out of exaggerated fears of remote disasters.

Note the underlying framework of this counterargument. The determinist premise is retained. The only difference is in the evaluation of the cost of environmental reform. We are still told we must choose between two variables, the industrial system and the environment. This is the framework of the trade-off theory that has emerged as the standard conservative response to environmentalism.

This theory pretends to be an application of economics, and some of its advocates are in fact economists. But the theory makes an incompetent application of the field it claims to represent, ignoring the dynamic character of economic development and the role of technology in periodically changing the terms of the economic equation. But economists do not intervene as energetically as they might to protest the abuse of their ideas in popular discourse. As a result, the trade-off theory plays a major role in politics and policy and so deserves serious discussion.

Despite their modern neoliberal dress, the conservative arguments go way back. They pose the dilemma that Mandeville mocked in a famous bit of doggerel at the end of the eighteenth century. In the preface to his poem, he denounced those silly enough to complain about the major environmental problem of his day, the filth of London's streets. In demanding cleanliness, they wish away the prosperity of the city, which is the cause of the filth. The poem concludes:

... Fools only strive To make a Great an honest Hive.
Bare Vertue can't make Nations live In Splendour;
they that would revive A Golden Age,
. must be as free For Acorns, as for Honesty.
(Mandeville 1970, 76)

Cost/benefit analysis of regulations is supposed to be able to precisely quantify and compare alternatives along the continuum that runs between Mandeville's "splendour" and a diet of acorns. For example, each incremental increase in the cleanliness of the air produces an incremental decrease in the number of respiratory illnesses. The policy choice is clarified by estimating the cost of tightening emission standards, then estimating the reduction in medical costs, and comparing the two figures.

But how credible are the results? There are enough problems with this approach to cast doubt on its claims, at least in general applications such as this. The current value we place on the various elements of trade-offs may not make much sense in scientific or human terms. Organizations tend to hide or exaggerate costs that might interfere with their plans, and it is difficult to know how to place a monetary value on such things as natural beauty and good health, but these values must be translated into monetary terms to enter the calculation. Trade-off arguments are thus often based on flimsy estimates of costs and benefits when they are not ideological expressions of hidden interests.

The main alternative is the imposition of environmental standards. Naturally, costs will come up in the debate over standards, but they will be evaluated more flexibly and alternative arrangements designed to deal with them discussed more freely if the issues are not boiled down to a pseudo-scientific calculation.

The question I will address in the rest of this chapter is whether cost/benefit analysis can supply an environmental philosophy. When so generalized, it has been used, along lines anticipated by Mandeville, to argue that too much environmentalism will end up impoverishing society. But do we really understand the issues when we start out from the notion that there are trade-offs between environmental and economic values? While there are obvious practical applications of cost/benefit analysis, I will argue that it fails as a basis for environmental philosophy. In this I agree with an extensive critical literature that focuses on the problem of quantification.[1] To this literature I will add a discussion of technological aspects of the trade-off approach.

I will argue that when applied not only locally to specific problems but also generally to civilizational projects such as environmental transformation, trade-offs imply technological determinism and the neutrality of technology. But these are principles of a philosophy of technology that has long since been superseded by more sophisticated approaches. Once that philosophy falls, the limits of cost/benefit analysis become apparent. I discuss these implications here in relation to historical examples introduced in chapter 1. In my conclusion I argue that environmentalism is not essentially about trade-offs. The question it poses concerns the kind of world we want to live in, not how much we can afford of this or that.

Background Assumptions

Economics is based on the proposition that multiple variables cannot be optimized at the same time. To optimize A, some of B must be sacrificed. While this seems obvious in daily life, it involves some questionable background assumptions in policy applications.

In the first place, the options in a trade-off must be clearly defined. But defined by whom? There is an unfortunate ambiguity on this point. The concept of trade-offs has an obvious source in common experience where the agent who chooses between the options also defines them. But when it is incorporated into economics, it borrows plausibility from that common experience while overstepping its limits. Economists can deploy technical tools that enable them to extend the notion of trade-offs to include purely theoretical alternatives that figure in no actual calculus of well-being. This may confuse the issues in public debate over live options.

Now, there are sometimes good reasons for the economists' extension of the concept, but it is important not to mix the ordinary and this technical sense of trade-offs. Most people would not consider the failure to earn income through prostitution as a trade-off of moral principles for money for the simple reason that prostitution is not a live option for them. Similarly, well-established environmental and safety standards are not up for grabs, and their theoretical cost, which may sound impressive, is irrelevant to present concerns.

There is a second assumption in the background of the trade-off approach. To make sense of talk about trade-offs, all other things must remain equal. This assumption is called *"ceteris paribus."* If laws change, if prices change, if the relation between goods changes, then it makes no sense to talk about trade-offs. *Ceteris paribus* may be plausible in some short-run economic decisions. When one composes a personal budget it is reasonable to assume that all other things will be equal, that one will not win the lottery or be struck by lightning or discover unexpected mutual dependencies between goods. But extend the time horizon to historical spans, and it is not at all plausible that things will remain equal. It is thus not surprising to find that the trade-off approach fails to explain cases such as the abolition of child labor that resemble contemporary environmental regulation. The changes involved cannot be understood on the model of a personal budget.

There is a good reason for this: *ceteris paribus* is confounded by cases in which pursuing one good unexpectedly makes it possible to obtain another competing good. In such felicitous cases what looks like a trade-off is something very different. This is a historical commonplace since obstacles to linear progress such as resource scarcities and regulation often lead to the emergence of new paths of development and new relations between goods. For example, the initial response of automobile makers to pollution controls reduced fuel efficiency, an undesirable trade-off. Later innovations culminating in electronic fuel injection successfully combined emission controls and fuel efficiency. Here, clearly, all things are not equal, and the trade-off dissolves in the face of technical advance.

Applied uncritically, *ceteris paribus* overlooks the possibility of such advances. Thus it implies that development proceeds along a fixed path from one stage to the next without the possibility of branching out in new directions inspired by regulatory interventions. Technological determinism hides in the background of this approach.

Deterministic applications of trade-off theory serve to challenge not only environmentalism but also many other technological reforms. For example, until recently most management theorists were convinced that there was a trade-off between worker participation and productivity. Technological imperatives supposedly condemned us to obedience at work (Shaiken 1984). Similar arguments in medicine keep patients in a passive role. In the early 1970s, women demanding changes in childbirth procedures were told they were endangering their own health and that of their babies. Today many of the most controversial changes have become routine, for example, partners admitted to labor and delivery rooms. When AIDS patients in the 1980s sought access to experimental treatment they were told they would impede progress toward a cure. Their interventions did not prevent the rapid discovery of the famous "drug cocktail" that keeps so many patients alive today (Feenberg 1995, chap. 5). Over and over technological reform is condemned as morally desirable perhaps, but impractical. Over and over the outcome belies the plausible arguments against reform.

Determinism is often accompanied by the belief in the neutrality of technology. As pure means, the only value to which technology conforms is the formal value of efficiency. The neutrality thesis is familiar from the gun-control debate where it is expressed in the slogan "Guns don't kill people, people kill people." Guns are neutral, and values are in the heads of those who choose the targets.

Together, technological determinism and the neutrality thesis support the idea that progress along the one and only possible line of advance depends exclusively on rational judgments about efficiency. Since only experts are qualified to make those judgments, environmentalists obstruct progress when they impose their "ideological" goals on the process of development. Where goals conflict, one or the other must be sacrificed: environmental protection or technological advance—in Mandeville's terms, virtue or prosperity.

The previous chapter presented an alternative view. I argued there that technological development can switch paths in response to constraints. On its new path, it may achieve several goals that were originally in conflict along its old one. Where the breakthrough to a new path responds to values articulated in the public sphere, a democratic technological revolution takes place.

This approach to technology is reminiscent of Thomas Kuhn's famous theory of scientific revolution. Kuhn showed that important scientific advances may appear purely rational, that is to say, uniquely determined by evidence and arguments, but they are actually underdetermined by rationality since they also respond to changes in the very idea of evidence and arguments (Kuhn 1962).

Technology is similar. The previous chapter discussed several examples. The regulation of child labor appeared to have unacceptable costs, but once put into effect it released new sources of wealth. The boiler code appears purely rational—surely a safer boiler is better from an engineering standpoint. But history shows that the decision to make safer boilers took forty years, and then the moving force behind the change was politics, not engineering. We thus have the same kind of problem in understanding the development of technology that Kuhn had with scientific development: progress is not reducible to a succession of rational choices because criteria of rationality are themselves in flux.

Kuhn's solution to this conundrum was the notion of paradigms, by which he meant a model for research. Such models have tremendous influence on those who come afterward. For example, physicists found in Newton not just a correct theory of gravitation but also a way to do physics that prevailed for several hundred years.

Normal science, Kuhn argued, is research within the established paradigm. The technological equivalent is the pursuit of efficiency in conformity with what I call "technical codes," the codes that govern technical practice (Feenberg 1999, 87–89). These codes materialize values in technical disciplines and design.

Revolutions in both science and technology involve fundamental changes in values reflected in the paradigms or codes that control the normal pursuit of truth or efficiency. Progress proceeds within a paradigm through the continuous advance of research and development, but there is discontinuity between paradigms. They open up incommensurable worlds.

This approach has consequences for our understanding of the rationality and autonomy of the technical professions. At every stage in the history of their discipline, experts inherit the results of earlier revolutions growing out of technical controversies and struggles. Engineering students do not have to learn how this or that regulation was translated into a design specification. The results are technically rational in themselves and presented as such. This gives rise to a characteristic illusion of autonomy. In fact the autonomy of these disciplines is limited. Their past is not a succession of decisions identifying the scientifically validated "one best way," but rather it is the result of social choice between several good ways with different social consequences. There is thus what might be called a "technological unconscious" in the background of these disciplines. This is what makes determinism so plausible, but it also leaves it vulnerable to historical refutation.

Two Historical Examples

In this section I will return to the earlier discussion of child labor and steam boilers in search of evidence for a nondeterministic position. Recall that Sir J. Graham, the opponent of labor regulation, believed that technological imperatives required the labor of women and children. There is a famous old photograph by Lewis Hine that helps to understand his concerns.[2] It shows a little girl in front of the equipment she uses in a cotton mill. She looks about ten years old, standing there in a white dress in front of ranks of machines going back into the distance. At first glance the picture seems quite ordinary. But soon one notices something strange about it: the machines are built to her height. The mill was designed for operation by children four feet tall. Industrial technology, like the chairs in an elementary school classroom, was designed for little people. The machines would be obsolete without children to operate them. Thus technological imperatives did indeed require child labor. The flaw in this argument is obvious today. Labor regulation resulted not in economic collapse but in the employment of more productive adults with machines adapted to their height.

Figure 2.1
"Girl Worker in Carolina Cotton Mill"
Library of Congress

Determinism misses the cultural dimension of this historical change. In developed countries, child labor violates fundamental assumptions about the nature of childhood. Today we see children as consumers, not as producers. Their function is to learn, insofar as they have any function at all, and not to earn a living. This change in the definition of childhood is the essential advance brought about by the regulation of labor.

In sum, although the abolition of child labor was promoted for ideological reasons, it was part of a larger process that redefined the direction of progress. In the child labor case, all other things were not equal because a new path of development emerged. On this path regulation actually contributed to increasing social wealth. Technology was not neutral in this case. It embodied the meaning of childhood in machines. This was a technological revolution.

The steamboat boiler case reveals another aspect of the problem. To us it seems obvious that regulation was needed. But apparently it was not obvious in the early nineteenth century. The situation was puzzling. Consumers kept on buying tickets despite the rising toll. At the same time, people voted for politicians who demanded regulation. It was reasonable to ask what people really wanted: cheap travel or safety. This ambiguity can be understood as a case of interpretive flexibility in the constructivist sense. Closure around the problem definition had yet to be achieved. But for there to be a trade-off account, the options must be stabilized. In the steamboat case the options were not stable. There were two slightly different and competing problem definitions: one at the individual and the other at the collective level, and it was not clear what the problem was.

The ambiguity was finally resolved, and the controversy settled once the problem was defined by an authoritative agent, the U.S. government, which prioritized the prevention of accidents. Of course no one was in favor of accidents, but their significance and the importance of preventing them depended on the context in which they were viewed.

In everyday life, our goals are nested in hierarchies. But sometimes particular actions or objects we pursue belong to several different hierarchies where they may have somewhat different meanings. In such cases an individual decision may well differ from a communal one because the community relates the options to different goals than do the individuals. Trade-offs are further complicated where these goals are associated with different decision procedures, each procedure introducing a different bias into the choice. This complication is relevant to the steamboat case. Individual market-based decisions led to different conclusions than collective political decisions because individuals and governments situate safety in different goal hierarchies.

Individual travelers simply wanted to reach their destinations cheaply. Like drivers who fail to fasten their seat belts today, they ignored the personal risk in their own individual case. But politics brought in other considerations besides personal risk. The basis for regulation is the commerce clause of the Constitution under which the government controls interstate transportation. This is not only a matter of economics but also of national unity. Like the highway system today, the canals and rivers of the early nineteenth century unified the territory of the United States. The movement of people, ideas, goods, troops—all the things that define a nation—depends on transport and in that period most especially on steamboats. National unity is not an individual economic concern but a collective political one. Safe transport had obvious individual benefits, and indeed most of the congressional debate concerned those benefits, but it was also a legitimate national issue. For example, senators from the West argued that they should not have to fear for their lives in traveling back and forth between the nation's capital and their constituents.

From an individual standpoint, the imposition of regulation traded off ticket prices for safety, but at the collective level something quite different was at stake. The infrastructure of national unity lies beyond the boundaries of the economy. It cannot be traded off for anything. Once security of transport is treated as essentially political, it ceases to figure in routine economic calculations. It no longer makes sense to worry about the slight increase in ticket prices once the principle of national interest in safe transportation is established. Just as we don't worry about all the money we are losing by not marketing our bodies for sex, so the cost of ensuring a certain minimum security of transportation figures in no one's account books.

Thus in this case the decision about what kind of technology to employ could not be made on the basis of efficiency for two reasons: First, because efficiency is relative to some known purpose. If the purpose is in question, efficiencies cannot be compared. And second, because efficiency is not relevant to questions of national unity.

Environmental Values

Now let me return to the question of the relation between environmental values and the economy with this constructivist argument in mind. I have identified several problems with the trade-off approach.

First, it ignores the significance of the shifting boundaries of the economy. We do not mourn the cost of using adult labor instead of child labor for the simple reason that children are culturally excluded from the category of workers.

Second, the trade-off approach assumes the fixity of the background, *ceteris paribus*, but technological change over the long time spans of history invalidates that assumption. All things are not equal in history since cultural change and technological advance alter the terms of the problem.

Third, the trade-off approach obscures differences in problem definition and goals reflecting different contexts of decision. There is no absolute context from the standpoint of which an unbiased evaluation is possible. It is thus deceptive to compare such things as the risk of death in an automobile accident with the risk of death from a nuclear accident since the one case involves individual responsibility and the other collective responsibility.

Fourth, the trade-off approach confuses short-run economic considerations with civilizational issues. These latter concern identity: who we are and how we want to live. This is a different proposition from getting more of A at the expense of B.

For all these reasons we need another way to think about environmental values. Here is a constructivist approach to an example that concerns a current environmental issue: the case of air pollution and asthma. Asthma attacks are treated as a cost in cost/benefit calculations. One study of the revised Clean Air Act valued asthma attacks at an average $32 (Rowe and Chestnut 1986). Obviously, the lower the cost of attacks, the less benefit is recovered by decreasing their frequency. Although calculations of this sort are offensive to anyone with asthma, it makes some kind of sense to the extent that our society is not fully committed to the struggle against this disease, which has modest medical costs.[3]

But it is entirely possible that we will respond to the rapidly rising incidence of asthma and the rising death rate associated with it by attempting to eliminate pollution as a causative factor. This would mean treating asthma the way we currently treat waterborne diseases such as cholera and dysentery. In that case, health-based standards would place asthma beyond the boundaries of economic controversy, and we would eventually arrive at a state of affairs that would seem obvious and necessary both technically and morally.

The relevant polluting methods would be replaced gradually by clean ones. Spare parts for the old polluting devices would be unavailable, and they would gradually go out of service if their use was not simply outlawed. After a while, the substitutes would be better in many respects, not just environmentally, since all later progress would be designed for them. It would not occur to our descendants to save money by going back to the old polluting machinery in order to cheapen industrial production or transportation. They would say, "We are not the kind of people who would trade off the health of our children for money," much as we would immediately reject the suggestion we supplement the family budget by sending our children out to work in a factory. This would be a civilizational advance in the environmental domain.

This leads to the question of why environmental values appear as values in the first place. Indeed, why is it at all plausible to claim that environmentalism is an ideology intruding on the economy? This is explained by the fact that our civilization was built by people indifferent to the environment. Environmental considerations were not included in earlier technical disciplines and codes, and so today they appear to come from outside the economy. It is this heritage of indifference that makes it necessary to formulate concern for the environment as a value and to impose regulation on industry.

This charge of indifference need not imply an overly harsh judgment of our predecessors. Not only are we richer and better able to afford environmental protection, but also the immense side effects of powerful technologies that have come into prominence since World War II have made environmental regulation imperative for us (Commoner 1971).[4] However, it does imply a harsh judgment of contemporaries who rely on specious arguments to justify blocking and dismantling regulations we can well afford today and desperately need. However powerful these conservative ideologues may look at the moment, we can expect their current offensive to fail as the severity of environmental problems makes an obvious mockery of their claims.

From this standpoint, it seems likely that the ideological form of environmental values is temporary. These values will be incorporated into technical disciplines and codes in a technological revolution we are living unawares today. Environmentalism will not impoverish our society. We will go on enriching ourselves, but our definition of prosperity and the technologies instrumental to it will change and become more rational in the future judgment of our descendants. They will accept environmentalism as a self-evident advance. Just as images of Dickens in the bootblack factory testify to the backwardness of his society, so will images of asthmatic children in smog-ridden cities appear to those who come after us.

What we have seen with child labor and safety standards is just as true of environmental standards. Once they are established, the old options drop out of sight. No one thinks about saving money by getting rid of seat belts in cars, and few car owners disable pollution control devices to improve performance. The only "trade-off" in which yesterday's bad designs play a role is in the head of conservative commentators. As zealous accountants they may insist that we monetize all these considerations and mark them down as expenses. But when the boundaries of the economy shift, so many cultural and technical consequences follow that it makes no sense to look back with an eye to costs and benefits. In the only sense in which, effects on social wealth are significant for policy they must be measured with respect to the fulfillment of actual desires, not theoretical constructions.

To be sure, we should be interested in economists' calculation of risks of which people are temporarily ignorant, such as the consequences of smoking. But that concerns a future in which live options can be expected to appear. Once the case is settled, the dead options are no longer relevant. And since it is impossible to put a price on revolutionary changes in the direction of progress, cost/benefit analysis can play only a minor role in such debates.

One might object that in failing to appreciate theoretical trade-offs, we ignore economic realities, but that is a short-term view. This type of cultural change is eventually locked in by technical developments.[5] For example, in the abstract one could redo all the calculations of labor costs taking into account the savings that might be made with cheap child labor, but that is an economic absurdity since developed economies presuppose the educated and disciplined products of schooling and could not be operated by children. Priorities change too, so it is impossible to compare the value of something like cleaner air or water with other goods on a constant basis over historical time.

It is thus a misrepresentation to claim that we are spending a specific sum such as $100 billion a year on environmental protection as though this money could be made available for other purposes. No doubt most of it went into improved design standards we now take for granted, for example, proper toxic waste disposal, safer water supplies, and so on. Economics regards these as "goods," and they do indeed have costs that may be controversial at first, but once they have been integrated to the culture and the prevailing technical environment, we do not think about those costs any more than New Yorkers conceive of Central Park as a piece of real estate they could sell if they wanted to buy something else for a change. In sum, economics can help us navigate the flow of wealth, but it cannot tell us where to place the dams that change the course of that flow.

Conclusion

Technological revolutions look irrational at first, but in fact they establish another framework of rationality, another paradigm. Thus it is neither rational nor irrational in some absolute sense to build a safer boiler. Constructivists would say that the decision to do so is "underdetermined" by pure considerations of technical efficiency because it also depends on a decision about the meaning of transportation and the significance of safety. As we have seen, that is a value question settled through political debate. Similarly, withdrawing children from the labor process and putting them in school was an enormous change, a change of civilization. Such a change is bound to generate a different path of technological development. With environmentalism we are again witnessing the opening of a new path.

Although its progress is slow and there are setbacks, environmentalism has the temporality of a revolution. Revolutions represent themselves as fully real in the future and criticize the present from that imagined outcome. The French revolutionary Saint-Just asked what "cold posterity" would someday have to say about monarchy even as he called for its abolition (Saint-Just 1968, 77). With history as our guide, we too can overleap the ideological obstacles to creating a better future by realizing environmental values in the technical and economic arrangements of our society.

3 Looking Forward, Looking Backward: The Changing Image of Technology

Utopia and Dystopia

In the year 1888, Edward Bellamy published a prophetic science fiction novel entitled *Looking Backward: 2000–1887* (Bellamy 1960). Bellamy's hero is a wealthy Bostonian who suffers from insomnia. He sleeps hypnotized in an underground chamber where he survives the fire that destroys his house. Undiscovered amid the ruins, he dozes on in suspended animation for more than a century, awakening finally in the year 2000 in a Boston transformed into a socialist utopia. Most of the book is taken up with his puzzled questions about his new surroundings and his hosts' lucid explanations of the workings of an ideal society.

Bellamy's book is now forgotten except by specialists, but it quickly became one of the bestsellers of all times, read by millions of Americans from the closing years of the nineteenth century until World War II. It articulated the hope in a rational society for several generations of readers.

In 1932, less than fifty years after Bellamy's famous book appeared, Aldous Huxley published *Brave New World*, a kind of refutation of *Looking Backward*. In the exergue to Huxley's book the Russian philosopher Berdiaeff regrets that "utopias appear to be far more realizable than used to be believed." Berdiaeff goes on: ". . . a new century is beginning, a century in which intellectuals and the cultivated classes will dream of the means of avoiding utopias and returning to a less 'perfect' and freer non-utopian society" (Huxley 1969). Unlike *Looking Backward*, *Brave New World* is still widely read. It is the model for many later "dystopias," fictions of totally rationalized societies in which, as Marshall McLuhan once put it, we humans become the "sex organs of the machine world" (McLuhan 1964, 46).

We can now literally "look backwards" at the twentieth century, and as we do so, the contrast between Bellamy's utopia and Huxley's dystopia is a useful one to stimulate reflection on what went wrong. And, clearly, something very important did go wrong to confound the reasonable hopes of men and women of the late nineteenth century. While they expected social and technical progress to proceed in parallel, in reality every major technical advance has been accompanied by catastrophes that call into question the very survival of the human race.

What happened to dash those hopes? Of course we are well aware of the big events of the century such as the two world wars, the concentration camps, the perversion of socialism in the Soviet Union, and more recently, the threats from genocidal hatreds, environmental pollution, and nuclear war. But underlying these frightful events and prospects, there must be some deeper failure that blocked the bright path to utopia so neatly traced by Bellamy.

Could a spiritual flaw in human nature or in modernity be responsible for the triumph of greed and violence in the twentieth century? No doubt human nature and modernity are flawed, but this is old news. Bellamy and his contemporaries knew all about the insatiable appetites, the pride and hatred lurking in the hearts of people. They understood the battle between Eros and Thanatos as much or as little as we do. What has changed is not our evaluation of human nature or modernity but the technical environment that has disrupted the delicate balance between the instincts that still left Bellamy's contemporaries room for hope, indeed for confident predictions of a better future.

We can begin to understand this technical shift by considering what is missing from Bellamy's description of society in the year 2000. His utopia is completely industrialized, with machines doing all the hardest work; improved technology and economies of scale have enriched the society. Workers are drafted into an "industrial army" where a combination of expert command and equal pay responds to the claims of technical necessity and morality. Although this is clearly an authoritarian conception, it is important to keep in mind that obedience is ethically motivated by the economic equivalent of patriotism rather than imposed through management techniques. Workers can freely choose their jobs after a brief period of manual labor at the end of regular schooling. The labor supply is matched voluntarily to demand by offering shorter workdays for less desirable jobs.[1]

Workers retire at forty-five and devote themselves to self-cultivation and to the duties of full citizenship that begin at retirement.

Bellamy's utopia is essentially collectivist, but paradoxically members of the society are depicted as highly differentiated individuals, each developing his or her own ideas, tastes, and talents in the generous allotment of leisure time made possible by technological advance. Individuality flourishes around the free choice of hobbies, newspapers, music and art, religion, democratic participation in government, and what we would call "continuing education." Invention too is an expression of individuality and a source of social dynamism.

None of these activities is organized by the industrial army because they have no scientific-technical basis, hence no technology requiring expert administration and no objective performance criteria. The economies of scale that make industrial technology so productive in Bellamy's account have no place in these activities, which depend on individual creativity.

Those who wish to act in the public sphere through journalism, religious propaganda, artistic production, or invention therefore withdraw from the industrial army as they accumulate sufficient "subscribers" to their services to justify the payment by the state of a regular worker's wage. The state provides these cultural creators with basic resources such as newsprint without regard for the content of their activities.

How different this imaginary socialism is from the real thing as it was established in the Soviet Union only a generation after Bellamy published his book! His society is bipolar, half organized by scientific-technical reason and half devoted to *Bildung*, the reflectively rational pursuit of freedom and individuality. But this bipolarity is precisely what did *not* happen in the twentieth century under either socialism or capitalism. Instead, total rationalization transformed the individuals into objects of technical control in every domain, and especially in everything touching on lifestyle and politics.

In Bellamy's vision, standardization and control were confined to the struggle with nature. Apparently, he could not conceive of what Norbert Wiener called "the human use of human beings." But the mass media continued the industrial pattern of efficiency through economies of scale. The twentieth century saw the displacement of higher culture in public awareness by a mass culture dedicated to unrestrained acquisitiveness and violent political passions.

It is interesting to see how close Bellamy came to anticipating mass society despite his blindness to the danger. At a time when telephone hook-ups still numbered in the thousands, he imagined a telephone-based broadcasting network that would disseminate preaching and musical performances. Each house would have a listening room, and programs would be announced in a regular printed guide. Bellamy foresaw that musical performance in the home would decline as broadcasts by professionals replaced it. So far his extrapolations are remarkably prescient, but nowhere did Bellamy anticipate the emergence of gigantic audiences subjected to commercial and political propaganda. Nor did he suspect that the small publications of his day, individual artistic production, and personal preaching would be so marginalized in the future that they would be unable to sustain individuality, which he took to be the ultimate goal of social life.

Brave New World, on the other hand, was written a decade after the first commercial radio broadcasts, which adumbrated a future of media manipulation. Huxley's vision was extrapolated from the rise of modern advertising, popular dictatorships, and mass production. In *Brave New World*, human beings are willing servants of a mechanical order. The Marxist hope, which Bellamy shares, for human mastery of technology no longer makes sense once human beings have themselves become mere cogs in the machine. This same view underlies much twentieth-century thought, for example, pessimistic social theories such as Max Weber's and the various deterministic philosophies of technology influenced by Martin Heidegger.

Dystopian Philosophy and Politics

Heidegger's philosophy of technology is a puzzling combination of romantic nostalgia for an idealized image of the premodern world and deep insight into modernity. His originality lies in treating technique not merely as a functional means but as a mode of "revealing" through which a "world" is shaped. "World" in Heidegger refers not to the sum of existent things but to an ordered and meaningful structure of experience. Such structures depend on basic practices characterizing societies and whole historical eras. These constitute an "opening" in which "Being" is revealed to human "*Dasein*," that is to say, in which experience takes place.

Modernity is characterized by what Heidegger calls "enframing." This concept describes a state of affairs in which everything without exception has become an object of technique. Things and people are now defined by their place in a methodically planned and controlled action system. For modern people everything is raw materials in technical processes, and nothing stands before being as the place of awareness. Complete meaninglessness threatens where the unique status of the human is so completely denied.

Heidegger might be thought of as the philosopher of *Brave New World*, except that he would deny that what we have before us today is a "world" in the full sense of the term. Rather, we are surrounded by an "objectless" heap of fungible stuff that includes us.

The Frankfurt School philosopher Herbert Marcuse was a student of Heidegger. His critique of "one-dimensional society" resembles his teacher's theory in a Marxist guise. Heidegger distinguished between craft labor, which brings out the "truth" of its materials, and modern technology, which incorporates its objects into its mechanism under the domination of a will and a plan. In Marcuse this Heideggerian approach continues as the distinction between the intrinsic potentialities that might be realized by an appropriate art or technique and the extrinsic values to which things are subordinated as raw materials in modern production. And like Heidegger, Marcuse deplores the extension of the latter approach from things to human beings.

But unlike Heidegger, Marcuse holds out the possibility in principle, if not much hope, of creating a new technology that respects the potentialities of human beings and nature. This "technology of liberation" would be a "product of a scientific imagination free to project and design the forms of a human universe without exploitation and toil" (Marcuse 1969, 19). This is still a worthy goal, although perhaps it should be described as a receding horizon: today we seem to be as far from achieving it as when Bellamy wrote.

These are what I call "dystopian philosophies of technology." I will have more to say about these philosophies in the third part of this book. They were surprisingly influential in the 1960s and 1970s despite their notorious pessimism and obscurity. Dystopian themes showed up not only in politics but also in films and other popular media, discrediting liberalism and gradually infiltrating conservatism. Contemporary politics is still strongly influenced by such vulgarized versions of the dystopian sensibility as distrust of "big government." These changes were accompanied by a dramatic shift in attitude toward technology. By the end of the 1960s, technophobia had largely replaced postwar enthusiasm for nuclear energy and the space program. No doubt the arrogance of the technocracy and the absurdity of the Vietnam War played a major role in this change.

Dystopian consciousness was transformed as it spread. No longer just a theoretical critique of modernity from the standpoint of a cultural elite, it inspired a populist movement. The question of technology was now a political question. The New Left reformulated socialist ideology in a tense synthesis between traditional Marxism and antidystopianism (Feenberg 1995, chap. 3).

The French May Events of 1968 was the high point of this remarkable change in the sensibility of the left. This was by far the most powerful New Left movement and the only one with massive working-class support. The May Events was an antitechnocratic movement, as hostile to Soviet-style socialism as to advanced capitalism. The students and militant workers proposed self-management as an alternative to planning and markets. No longer cogs in the machine, they demanded freedom. Their position was summed up in a widely circulated leaflet: "Progress will be what we want it to be" (Feenberg and Freedman 2001, 84).

The May Events was one of many similar movements that challenged the conventional idea of progress. These movements opened a space for the new technical politics of recent decades that engages in concrete struggles in domains such as computers, medicine, and the environment.

The Impact of the Internet

Although the movements of the 1960s undermined technological determinism both in theory and practice, they continued to employ a dystopian rhetoric in response to the technocratic threat. However, as the twentieth century came to a close, dystopianism lost much of its authority, and utopia returned in a new guise. Technology plays a central role for Bellamy and Huxley, but the advances they describe are symbols of hopeful or disastrous social trends rather than specific technological forecasts. Contemporary utopias are presented as breathless frontline reports on the latest R and D. These new utopias are inhabited by bioengineered "transhumans" enhanced by drugs, networked in a universal mind or downloaded to more durable hardware than the human body. Determinism returns as social consequences are deduced from future technology. Serious thinkers perplexed by this upsurge of horrific speculation once again raise flimsy ethical barriers to "progress." Antidystopian humanism struggles to salvage spirit from the satanic mills of advancing technology. But now the whole contest seems routine and not very credible.

Meanwhile, new and more interesting trends have emerged among researchers who eschew speculation and study technology as a social phenomenon. These researchers view the dystopian critique of modernity and humanistic ideology as nostalgic longing for a past that is forever lost and that was not so great in any case. According to this view, we belong wholly and completely to the technological network and do not represent, nor should we await, a suppressed alternative in which "man" or *"Dasein"* would achieve recognition independent of his tools.

Nonmodern or posthumanist thinkers such as Bruno Latour and Donna Haraway have put forward this approach with singular energy in books and essays with titles such as *We Have Never Been Modern* (Latour 1993) and "The Cyborg Manifesto" (Haraway 1991). The very tone of these titles announces an agenda for the new millennium. According to the authors, we have passed through the experience of dystopia and come out on the other side. Our involvement with technology is now the unsurpassable horizon of our being. No longer opposed to technology, we join together with it in a more or less undifferentiated "cyborg" self. It is time to cease rearguard resistance and, embracing technology once and for all, give its further development a benign direction.

The Internet supplies the essential social background to the wide interest in this posthumanist view. Of course the authors did not have to go online to develop their ideas, but the popular credibility of their innovative vision depends on the emergence of computer networking and the new function of subjectivity it institutes. Without the widespread experience of computer interaction, it is unlikely that their influence would have spread beyond a narrow circle of researchers in science studies. But given that experience, they articulate a fundamental shift from antagonism to collaboration in the relation of human beings to machines.

What is it about networking that assuages dystopian consciousness? The fear of dystopia arises from the experience of large-scale social organization that, under modern conditions, possesses an alienating appearance of rationality. Technocratic domination was exemplified in the mass media audiences of the twentieth century until computer networking broke the pattern. Instead of the passivity associated with broadcasting, the online subject is constantly solicited to "interact" either by making choices or responding to communications. This interactive relationship to the medium, and through it to other users, appears nonhierarchical and liberating. Like the automobile, that fetish of modernity, the Internet opens rather than closes vistas. But unlike the automobile, the Internet does not merely transport individuals from one location to another; rather it constitutes a "virtual" world in which the logic of action is participative and individual initiative supported rather than suppressed by technology.

It is noteworthy that this evolution owes more to users than to the original designers of the network, who intended only to streamline time sharing and the distribution of information. Refuting technological determinism in practice, users "interacted" with the network to enhance its communicative potential. This was the real "revolution" of the "Information Age" that transformed the Internet into a medium for personal communication. As such it resembles the telephone network in which the corporate giants who manage the communication have little or no control over what is communicated. Such systems, called "common carriers," extend freedom of assembly and so are inherently liberating.

What is more, because computer networking supports group communication, the Internet can host a wide variety of social activities, from work to education to exchanges about hobbies and the pursuit of dating partners. These social activities take place in virtual worlds built with words. The "written world" of the Internet is indeed a place where humans and machines appear to be reconciled (Feenberg, 1989b).

At this point, a note of caution is in order. The enthusiastic discourse of the Information Highway has become predictable and tedious. It awakens instant and to some extent justified skepticism. It is unlikely that the twenty-first century will realize the dream of a perfectly transparent, libertarian society in which all people can work from their home, publish their own book, choose multiple identities and genders, find a life partner online, buy personalized goods at an electronic mall, and complete their college education in their spare time on their personal computer. It is reasonable to be suspicious of this vision. The dystopian critic finds here merely a more refined and disguised incorporation of the individual into the system.

But both utopian and dystopian visions are exaggerated. The Internet will certainly have an impact on society, but it is ludicrous to compare it with the Industrial Revolution, which pulled nearly everyone off the farm and landed these people in a radically different urban environment. My "migration" to cyberspace over the last thirty years can hardly be compared with my grandparents' migration from a central European village to New York. Worrisome though it may be, the "digital divide" is far more easily bridged than the divide between city and country in a society without telephones, televisions, and automobiles. Unless something far more innovative than the Internet comes along, the twenty-first century will be a continuation of the twentieth, not a radical and disruptive break. The real significance of the Internet lies not in the inauguration of a new era but in the smaller social and technological changes it makes possible within a largely familiar framework.[2]

New Forms of Agency

The most important of these changes has to do with democratic agency. Technical communities have been able to use the Internet to coordinate their demands for a fuller representation of their interests. Despite discouraging developments in other domains, agency in the technical sphere is on the rise. The new online politics will not replace electoral politics, but its existence has extended the public sphere to embrace technical issues formerly considered neutral and given over to experts to decide without consultation. This has had the effect of creating a social and technical environment in which agency in the traditional domain of politics has begun to recover from the passivity induced by a steady diet of broadcasting.

There are many examples of online political activism: the use of the Internet by the Zapatista movement in Mexico, protests against the WTO and the IMF, opposition to the Iraq War, Howard Dean's and Barack Obama's campaigns, and many other similar interventions. The Internet has broken the near-monopoly of the business- and government-dominated official press and television networks by enabling activists to organize and to speak directly to millions of Internet users (McCaughey and Ayers 2003; Milberry 2007).

But agency online is not confined to politics, and this fact is important for evaluating its significance. After all, these political examples might be odd exceptions and the Internet principally defined as an electronic mall, as its critics charge. I believe on the contrary that political usages of the Internet are instances of a much broader phenomenon, the emergence of new forms of agency in online communities of all sorts (Bakardjieva 2009).

The earliest movement of this sort formed on proprietary networks before the public had access to the Internet, although it has used the Internet to greatly amplify its scope and influence. Software users form an invisible community that has until recently been helpless before gigantic firms such as Microsoft that are notoriously indifferent to users' demands. But the software business is young. In the early days of the IBM mainframe, users rather than commercial suppliers developed software. Habits of free exchange acquired then gradually merged with an ideological movement for free and open source software initiated by Richard Stallman and many others. The rapid development of the field thereafter has had a huge impact on the Internet. Each open source project gathers an online community that tests the programs and suggests or actually codes improvements. Software users and producers are no longer separated by the barrier of commercial enterprise but, like readers and writers in other types of online forums, can exchange places and engage reciprocally with each other.

Medicine is another domain of what Maria Bakardjieva calls "subactivism." Patients gather online to support each other, to share advice, and to make demands on the medical community. In 1995 I studied an early example, a discussion forum for patients with the rare neurologic disease ALS (amyotrophic lateral sclerosis, or Lou Gehrig's disease). The patients exchanged social support, lore about living with the disease, and information about medical experimentation. This new type of patient organization defied standard assumptions about the sick role. Instead of waiting in isolation for individual help from the medical profession, the patients worked together to further their interests. They eventually brought pressure to bear on the ALS Society of America to demand larger budgets and changed policies from the National Institutes of Health. Today similar patient forums proliferate on the Internet and create a very different social environment for medicine (Feenberg et al. 1996).

Video games offer a surprising example of user initiative. The video game industry is now larger than Hollywood and engages millions of subscribers in online multiplayer games. The players' gaming activities are structured by the game itself, but online communities organize them in informal relationships that the industry does not control. These online forums are venues for various unexpected appropriations of the game environment. For example, players auction game items for real money. Hackers have modified games, and the modified versions have occasionally become popular. Legal issues arise in such cases since players usually agree to extremely restrictive policies when they subscribe. Companies have generally responded angrily to violations at first, but in most cases they soon ignore the violators or modify their policies to accommodate them. The online game world thus supports a certain degree of interaction between customers and suppliers, different from what we have come to expect from television and film (Grimes 2006; Grimes and Feenberg 2009; Kirkpatrick 2004).

It would be easy to multiply examples. The academic world is especially active. For example, libraries have struggled to redefine their role as information providers in the face of competition from the Internet. They have begun to cross the line between stocking and publishing information. Many now support the creation of open access online journals in an effort to fulfill their traditional functions as noncommercial information brokers. Scholarly communities that formerly depended on the costly services of publishers can now organize themselves with the help of libraries (Willinsky 2006).

These and many similar instances of agency on the Internet situate online political activism in its context. We are witnessing the end of dystopia as the defining technology of our time shifts from great centralized systems such as electric power and broadcasting to the more loosely structured world of the computer.

Democratic Interventions

Langdon Winner describes technology as a kind of constitution insofar as it determines the framework of our lives and decides important political questions in the shape it gives our social relations as we use it (Winner, 1986: 47ff). Given its variety, technology would perhaps be more accurately compared with a code of laws. And like legislation, technology serves the interests and concerns of some better than others. Just as it is possible to trace out the links between laws and those they represent, so technologies can be said to represent their users. This is a reason to prefer a democratic technological regime that, like political democracy, enables the broadest possible representation.

But there are also important differences between politics and technology. The idea of representation is traditionally tied to geographical locality on the presumption that those who live close together share common interests and are able to meet to discuss them. Of course there are likely to be disagreements, but so long as communication is possible, conflicts can be resolved by legitimate means such as voting.

Yet as we move into a more advanced phase of technological development, this rather narrow definition of politics inherited from the preindustrial past is less and less plausible. More and more aspects of social life are conditioned by commonalities among people who share a similar relation to the vast technical systems that shape social life. Technologically advanced societies enroll their members in a wide variety of technical networks that define careers, education, leisure, medical care, communication, and life environments. These networks are administered by experts and managers rather than democratically. They overlay the geographical communities and compete with them in significance in the lives of citizens. Shared concerns I call "participant interests" arise in this context (Feenberg 1999, chap. 6).

Obtaining adequate representation of these interests was well beyond the means of almost all technically organized populations in the days before the Internet. Only groups organized around politics in the traditional sense were also able to function effectively as technical pressure groups. The labor movement, for example, was able to impress governments with the importance of health and safety rules for industry. The movement for gay rights was able to influence the health system with demands for access to experimental AIDS drugs. But most participants in technical networks went unmobilized, and it appeared that technological advance would culminate in a technocratic order.

Already in the 1920s John Dewey foresaw the problems that would result. He worried that traditional local community was losing its integrity in a mobile modern society. New forms of technically mediated community were needed to replace or supplement localism, but these were not easy to create. The new links being forged by the advancing technical system were still inarticulate. Dewey remained caught in the dilemma he so presciently identified—large-scale technical networks as the form of modern social life and local community as the site of true democratic deliberation (Dewey 1980, 126).

This contradictory dualism of a modern society is well captured by Habermas's distinction of system and lifeworld (1984, 1987). Habermas analyzes markets and administrations as "systems" that coordinate social action objectively. The potentially conflicting intentions of the individuals are harmonized not by explicit agreements but by a technically rational institutional framework and simple procedural rules. Buyers and sellers, for example, act together on the market for their own mutual benefit with a minimum of cooperation. They need only recognize the forms of exchange such as price, purchase, and sale. Systems make possible the large-scale social organization of modern societies.

In contrast, the lifeworld consists of communicating subjects whose action is coordinated by mutual understanding of a wide variety of elaborate social codes and meanings. Production is organized primarily through the system, and social reproduction through the lifeworld. The dystopian critique of modernity can be reformulated on these terms as the growing predominance of system over lifeworld, with potentially disastrous consequences for social cohesion and the survival of individuality.

Despite the fruitfulness of Habermas's framework, it is fraught with problems. His schema leaves out technology, even though it too coordinates action objectively. Furthermore, he wavers between treating his concepts of system and lifeworld as pure analytic categories, cutting across all institutions and activities, and identifying them with specific institutions such as the market and the family. What is lost as a result of these omissions and ambiguities is a sense of the complexity of the real interactions between system and lifeworld.

Here is an example from the realm of communication technology. The Internet is a system in Habermas's sense, managed in accordance with administrative rationality and distributed on a market. As such it supports agencies and corporations with tremendous political and economic power. Yet the activities the Internet mediates are essentially communicative. In the lifeworld the Internet takes on meanings and connotations having to do with intimacy, human contact, self-presentation, creativity, and so on. The Internet is not merely instrumental to these lifeworldly ends; it belongs to the lifeworld itself as a richly signified artifact. This is more than a matter of subjective associations since it affects the evolution and design of the network and the interface, which cannot be understood in terms of an abstract idea of efficiency. This has become clear with the struggle over network neutrality. The intertwining of function and meaning exemplified by the Internet is general in modern societies.

The evolution of technical design reveals the agency of those treated as objects of management in the dominant technical code. The distinction of system and lifeworld appears to break down. But it would be a mistake to abandon it altogether. "Systems thinking" is not the exclusive prerogative of the social critic but rather grows out of the actual experience of managing modern organizations. Similarly, the lifeworld is not just an analytic category or a separate sphere such as the family but a perspective that is brought to bear on systems by those enrolled within them. As such it reflects the way in which alienation is lived and resisted by subordinate technical actors.

Michel de Certeau introduced an approach to the interaction of system and lifeworld that both preserves the essence of the distinction and grasps interactions that appear anomalous in Habermas's framework. I have applied this approach to technology. De Certeau distinguishes between the *strategies* of groups with an institutional base from which to exercise power and the *tactics* of those who are subject to that power and who, lacking a base for acting continuously and legitimately, maneuver and improvise micropolitical resistances (de Certeau 1980). The strategic standpoint privileges control and efficiency, while the tactical standpoint gives meaning to the flow of experience shaped by strategies. In the everyday lifeworld masses of individuals improvise and resist as they come up against the limitations of the technical systems in which they are enrolled. These resistances influence the future design of the systems and their products.

Interpreting the system/lifeworld relation in terms of strategies and tactics goes beyond both dystopian condemnation and the uncritical transhumanist celebration of technology. Dystopianism adopts the strategic standpoint on technology while condemning it. Technology is conceived exclusively as a system of control, and its role in the lifeworld is overlooked. This is why resistance appears impossible or impotent from this standpoint. But the introduction of a distinction between system and lifeworld also corrects transhumanism's overly optimistic picture of technology by recalling the role of meaning and human relations in modern technological systems. The contradiction between the system and the life-world of its users and victims explains the rise of struggles on the Internet in the emerging technical public sphere.

Conclusion

The Internet supports a vision of harmonious coexistence between humans and their machines. But these theoretical considerations pinpoint something rather different that was well understood by dystopian thinkers. They argued that technology is a source of power over human beings and not merely an instrument for the satisfaction of human needs. Because that power is essentially impersonal, governed by technically rational procedures rather than whims or even interests in the usual sense of the term, it appears to lie beyond good and evil. This is its dystopian aspect.

What Marcuse called "one-dimensionality" results from the disappearance of external agents of change and their transcending critique. But the exercise of technical power evokes resistances immanent to one-dimensional society. Technological advance unleashes social tensions whenever it slights human and natural needs. Because the system is not a self-contained expression of pure technical rationality but emerged from two centuries of deskilling and abuse of the environment, such slights occur often. Vocal technical publics arise around the resulting problems. Demands for change reflect aspects of human and natural being denied by the technical code of the system. The Internet provides a scene on which dystopia is overcome in a democratizing movement the full extent of which we cannot yet measure.

The utopian and dystopian visions of the late nineteenth and early twentieth centuries were attempts to understand the fate of humanity in a radically new kind of society in which most social relations are mediated by technology. The hope that such mediation would enrich society while sparing human beings themselves was disappointed. The utopians expected society to control modern technology just as individuals control traditional tools, but we have long since reached the point beyond which technology overtakes the controllers. But the dystopians did not anticipate that once inside the machine, human beings would gain new powers they would use to change the system that dominates them. We can observe the faint beginnings of such a politics of technology today. How far it will be able to develop is less a matter for prediction than for practice.

II Critical Constructivism

The first chapter of this part combines insights from philosophy of technology and constructivist technology studies in a critical theory of technology. Technologies are analyzed at two levels that cut across every device and system. At the primary level, people and things are decontextualized to identify affordances. Although essential to everything technological, the primary level by itself does not suffice to constitute a technology. At the secondary level, the decontextualized elements are recontextualized to fit in with their natural, technical, and social environments. This recontextualization process is also essential. The technical code is the rule under which technologies are realized in a social context with biases reflecting the unequal distribution of social power. Subordinate groups may challenge the technical code, influencing the evolution of technical design.

The second chapter applies these concepts to the first successful domestic computer network, the French Minitel system. The system was designed to distribute information in accordance with predictions of a postindustrial "information age," but it grew into something quite unexpected as users redirected it toward human communication. The first major computer network thus deviates sharply from the theories that were its original raison d'être. A closer look at this case shows the role of user agency in countering the technocratic bias of the dominant conception of postindustrial society.

The third chapter focuses on Japan, the first non-Western country to modernize. Japan represents a test case for the universality of modern achievements. This chapter shows the relevance of the Japanese experience through an analysis of several examples of technology transfer and through a discussion of Kitaro Nishida, the major prewar Japanese philosopher. The chapter introduces the concepts of "branching" and "layered" technological development and applies them to Nishida's theory of "place" (*basho*). The Japanese case does not resolve our questions about the nature of modernity, but it does show that technological rationality is culturally relative in complex ways. We must globalize our conception of technology, which can no longer be identified exclusively with Western achievements.

4 Critical Theory of Technology: An Overview

Technology and Culture

In standard accounts of technology, efficiency serves as the principle of selection between successful and failed technical initiatives. Because efficiency is a calculable quantity, technology appears to borrow the virtues of necessity and universality generally attributed to scientific rationality. Critical theory of technology demystifies this image by showing that technology is not merely instrumental to specific goals but shapes a way of life. That wider impact may be intended or unintended; it may result from specific design choices or from side effects. In any case the impact of technology is not a quantity but a quality and has nothing to do with universal rationality. It requires a very different kind of explanation.

Constructivist sociology of technology shows that different configurations of resources can yield alternative versions of the same basic device capable of efficiently fulfilling its function. The different interests of the various actors involved in design are reflected in subtle differences in the function and side effects of what is nominally the same device. Efficiency is thus not decisive in explaining the success or failure of similar designs since several viable options usually compete at the inception of a line of development. Technology is "underdetermined" by the criterion of efficiency and responsive to the various particular interests and ideologies that select among these options. Social choices intervene in the problem definition as well as its solution. Technology is not "rational" in the old positivist sense of the term but socially relative. This explains how the outcome of technical choices can be a world that supports the way of life of one or another influential social group.

I have introduced the concept of "technical code" to articulate the relationship between the social and technical design (Feenberg 2002, 74–80). A technical code is the realization of an interest or ideology in a technically coherent solution to a problem. Although some technical codes are formulated explicitly by technologists themselves, the term as I use it refers to a more general analytic tool that can be applied even in the absence of such formulations. More precisely, then, a technical code is a criterion that selects between alternative feasible technical designs in terms of a social goal and realizes that goal in design. "Feasible" here means technically workable. Goals are "coded" in the sense of ranking items as ethically permitted or forbidden, aesthetically better or worse, or more or less socially desirable. "Socially desirable" refers not to some universal criterion but to a widely valued good such as health or profit. Technical codes are formulated by the social theorist in ideal-typical terms, that is, as a simple rule or criterion. A prime example in the history of industrialization is the imperative requirement to deskill labor through mechanization rather than preserving or enhancing skills.

In every case, a technical code describes the congruence of a social demand and a technical specification. It is generally materialized in two different ontological registers: discursive and technical. A process of translation links the two. For example, demand for greater attention to automotive safety is translated into seat belts and air bags; operationally speaking, these functionalizations are what safety *means*. Thus technology and society are not alien realms as are facts and values in the treatises of philosophers. Rather they communicate constantly through the realization of values in design and the impact of design on values. This fluidity of the technical, highlighted in Bruno Latour's concept of delegation, explains why the vaunted trade-off of efficiency and ideology, dear to conservative business commentators, is largely mythical (Latour 1992).

Technical codes are always biased to some extent by the values of the dominant actors. The critical theory of technology aims to uncover these biases. Technical bias, however, is difficult to identify since the unjust social consequences of technical decisions appear to be mere side effects of "progress." Where technical codes are reinforced by individuals' perceived self-interest and law, their political import usually passes unnoticed. This is what it means to call a certain way of life "culturally secured" and a corresponding power "hegemonic."

Just such a hegemonic power is supported by the liberal notion that democratic capitalism is a value-neutral system in which everyone can pursue his or her private conception of the good. To show that the system is inherently biased requires an unfamiliar type of argument that has most often been deployed by certain philosophers of technology. They reject the alternative—technical rationality or social bias—and argue that the latter shows up in the former through the social content of technical choices. Examples of such arguments are Marcuse's notion that the neutrality of technology places it in the service of the dominant social groups and Albert Borgmann's critique of the mutual implication of liberalism and the "device paradigm" in a culture biased toward private consumption (Marcuse 1964; Borgmann 1984).

Critical theory of technology generalizes such arguments through a distinction between types of bias (Feenberg 2002, 80ff; see also chap. 8). The usual commonsense notion of bias attributes unjust discrimination to prejudice and emotion. This "substantive bias" is based on factually questionable beliefs. Substantively biased decisions in the technological realm, where cool rationality ought to prevail, lead to avoidable inefficiencies and breakdowns. But efficient operations are often unfair even where bias in this ordinary sense is avoided. Thus critical theory of technology introduces the concept of "formal bias" to understand how a rationally coherent, well designed, and properly operated technical device or system can nevertheless discriminate in a given social context. The concept of formal bias also sheds light on notions such as institutional racism and serves much the same purpose, namely, to enable a critique of socially rational activities that appear fair when abstracted from their context but have discriminatory consequences in that context. Today justice requires identifying and changing formally biased technical codes.

Operational Autonomy

For many critics of technological society, Marx is now irrelevant, an advocate of outdated economic theories. But Marx had important insights for philosophy of technology that must not be lost along with his discredited economics. He focused so exclusively on economics because production was the principal domain of application of technology in his time. With the penetration of technical mediation into every sphere of social life, contradictions and potentials similar to those he identified in the factory follow as well.

In Marx's view, the capitalist is ultimately distinguished not just by ownership of wealth but also by control of the conditions of labor. The owner has a technical as well as an economic interest in what goes on within his factory. By reorganizing the work process, he can increase production and profits. Control of the work process, in turn, gives rise to new ideas for machinery, and the mechanization of industry follows in short order.

This leads over time to the invention of a specific type of machinery that deskills workers and requires management. Management acts technically on persons, extending the hierarchy of technical subject and object into human relations in pursuit of efficiency. Eventually professional managers represent and in some sense replace owners in control of the new industrial organizations. Marx calls this the "impersonal domination" inherent in capitalism in contradistinction to the personal domination of earlier social formations. It is materialized in the design of machines and the organization of production. In a final stage, which Marx did not anticipate, techniques of management and organization and types of technology first applied to the private sector are exported to the public sector, where they influence government administration, medicine, and education. The whole life environment of society comes under the rule of technique. In this form the technological essence of the capitalist system can be transferred to socialist regimes built on the model of the Soviet Union.

The entire development of modern societies is thus marked by the paradigm of unqualified control over the labor process on which capitalist industrialism rests. Technical development is oriented toward the disempowering of workers and the massification of the public. This is "operational autonomy," the freedom of the owner or her representative to make independent decisions about how to carry on the business of the organization, regardless of the views or interests of subordinate actors and the surrounding community. The operational autonomy of management and administration positions them in a technical relation to the world, safe from the consequences of their own actions. These consequences may be dire where the enterprise rides roughshod over worker and community interests, but from the suppression of the Luddites down to the present, the agents of enterprise have usually been protected from the resulting outcry. In addition, operational autonomy enables them to reproduce the conditions of their own supremacy at each change in the technologies they command. Technocracy is an extension of such a system to society as a whole in response to the spread of technology and management to every sector of social life. Technocracy armors itself against public pressures, sacrifices community values, and ignores needs incompatible with its own reproduction and the perpetuation of its technical traditions.

The technocratic tendency of modern societies represents one possible path of development, a path shaped by the demands of power. In subjecting human beings to technical control at the expense of traditional modes of life while sharply restricting participation in design, technocracy perpetuates elite power structures inherited from the past in technically rational forms. In the process it mutilates not just human beings and nature but also technology. Technology has beneficial potentialities that are suppressed under capitalism and state socialism. These potentialities could be realized along a different developmental path were power more equally distributed.

Critical theory of technology identifies the limits of the technical codes elaborated under the rule of operational autonomy. The very same process in which capitalists and technocrats were freed to make technical decisions without regard for the needs of workers and communities generated a wealth of new "values," ethical demands forced to seek voice discursively. Democratization of technology is about finding new ways of privileging these excluded values and realizing them in technical arrangements.

A fuller realization of technology is possible and necessary. We are more and more frequently alerted to this necessity by the threatening side effects of technological advance. These side effects constitute feedback loops from the objects of our technical control to us as the subjects of that control. Normally the feedback is reduced or deferred so that the subject of technical action is safe from the power unleashed by its own actions. But technology can "bite back," as Edward Tenner reminds us, with fearful consequences as the feedback loops that join technical subject and object become more obtrusive (Tenner 1996). Today we are most obviously aware of this from the example of climate change, an unintended consequence of almost everything we do. The very success of our technology ensures that these loops will grow shorter as we disturb nature more violently in attempting to control it. In a society such as ours, which is completely organized around ever-more-powerful technologies, the threat to survival is clear.

Instrumentalization Theory[1]

Although critical theory of technology seeks to identify social aspects of technology, this approach does not preclude recognition of the importance of simple functionality. Technologies must really "work" to serve in social strategies, and the one desideratum cannot be reduced to the other.

Accordingly, critical theory of technology distinguishes analytically between the aspect of technology stemming from the functional relation to reality, which I call the "primary instrumentalization," and the aspect stemming from its social involvements and implementation, which I call the "secondary instrumentalization." Together these two aspects of technology constitute the "world" in something like the sense Heidegger gives the term.[2]

The primary instrumentalization initiates the process of world making by de-worlding its objects in order to reveal affordances. It tears them out of their original contexts and exposes them to analysis and manipulation while positioning the technical subject for distanced control. De-worlding reduces elements of reality that can be functionally deployed to their functional characteristics and situates them in a new context where they serve a purpose of some sort.

The theory is complicated, however, by the fact that technical devices and systems are built up from simple elements that have a wide variety of potentialities. The process in which these elements are combined consists in the successive decontextualizations and recontextualizations through which more and more limitations are imposed on the materials. The primary instrumentalization is iterative: individual elements configured to serve a given purpose can be further decontextualized and joined to other elements in combinations serving other purposes. For example, a rock may be picked up and used to crack open a shell. In the process it loses its connections to its original surroundings. But the same rock may later be stripped of its new function and recontextualized as a component of a larger entity formed by attaching it to a stick to make a hammer useful for construction work.

Social choices contribute to determining the specifics of the process. Here is a simple example: a tree is cut down and stripped of its branches and bark to be cut into lumber. All its connections to other elements of nature except those relevant to its place in construction are eliminated. But this aspect of the production process is not purely negative. It incorporates many socially motivated decisions. For example, such elementary specifications as the standard width of wooden boards differ from one country to another. Nothing can be done with the tree until that specification is determined. This process of social determination is the "secondary instrumentalization." It establishes the social meaning of the artifact.

At the secondary level, technical objects are integrated with each other as the basis of a way of life. The primary level simplifies objects for incorporation into a device, while the secondary level integrates the simplified objects to a social environment. This involves a process that, again following Heidegger, I will call the "disclosure" or "revealing" of a world. Disclosure qualifies the original functionalization by orienting it toward the world it contributes to creating.

The social appears in the technical domain in two principal forms I call "systematizations" and "valuative mediations." "Systematization" refers to the system of socially established meanings that determines the nature of technologies and the interconnections between their various parts and their technical, human, and natural environments. Since there is no unique causal logic determining the optimum interconnections, empirical study finds social choice in this technical aspect. Valuative mediations govern aspects of technologies that fall under ethical, aesthetic criteria and other general social norms. These aspects are not limited to prohibitions and external appearances but penetrate to the technical heart of the object. They form the cultural horizon of the society as it shapes technical artifacts.

Automobiles exemplify both aspects: they are designed for private ownership and use and work with specific types of roads and fuels (systematization); stylistically they appeal to various aesthetic tastes (mediation), these latter influencing in turn technical features such as dimensions and engine position. The interaction of these two aspects with automotive engineering is an iterative process in which the meaning that technologies take on in the lifeworld feeds back into their design from one stage in their development to the next.

The evolution of the refrigerator illustrates the relativity of design to both meaning and cultural horizon as exemplified by gender and environmental considerations. The standard volume specification reflects family size, while the outer form of the refrigerator was "streamlined" in the 1930s for incorporation into the space of women's domestic work (Nickles 2002).[3] At a later stage the discovery that the refrigerant was destroying the ozone layer led to a redesign responsive to environmental concerns. No device is too banal for the social study of technology.

It is important to keep in mind that the two instrumentalizations are only analytically distinguished in most cases. No matter how abstract the affordances identified at the primary level, they carry social content from the secondary level in their approach to the materials. Similarly, secondary instrumentalizations such as aesthetic design presuppose the identification of the affordances to be assembled and concretized. Cutting down a tree to make lumber and building a house with it are not the primary and secondary instrumentalizations, respectively. As I mentioned previously, cutting down a tree "decontextualizes" it, but in line with various specifications determining the size and shape of boards. Furthermore, technical, legal, and aesthetic considerations determine what kinds of trees can become lumber. The act of cutting down the tree is thus not simply "primary" but involves both levels as one would expect of an analytic distinction.

The impact of the secondary instrumentalizations increases as we follow an artifact from its earliest beginnings through the successive stages of its development into a finished device. The logging stage: the tree is cut down, but only the legal tree. The processing stage: it is transformed into lumber in accordance with the specifications of a particular construction system. The building stage: the house is built out of the lumber according to a building code and an architectural aesthetic. Even after its release, a technical device is still subject to further transformations through user initiative and government regulation: houses are remodeled. The lifeworld in which artifacts originate and to which they return has the power to shape and modify them. In this limited sense we can say they are socially shaped or "constructed."

The two instrumentalizations characterize technical production in all societies but are clearly distinguishable only in modern times. This has led to the illusion that they are entirely separate, externally related processes.

In fact the distinction is primarily analytic even today, although large organizations often separate certain social functions, such as packaging, from engineering operations. Thus the aesthetic function, an important secondary instrumentalization, may be separated out and assigned to a corporate "design division." Artists will then work in parallel with engineers. This partial institutional separation of the instrumentalizations encourages the belief that they are completely distinct. The existence of technical disciplines appears to confirm the commonsense notion that technology and society are separate entities, but these disciplines show traces of past social choices that have been crystallized in standards and materials. A technological unconscious masks this history.

Nevertheless, radical versions of constructivism are wrong to insist that there is literally no distinction between the social and the technical. If that were true, there would be no technical disciplines, and the makers and users of far simpler products would communicate more easily. It would be more accurate to say that modern technology is a particular expression of the social in artifacts and systems, mediated by differentiated technical disciplines. Ordinary social belief and behavior are quite different, mixing the technical and nontechnical promiscuously. Meanings guide improvisational action in everyday life, forming patterns that intersect with difficulty with engineered products, as Lucy Suchman argues persuasively (Suchman 2007).

An adequate philosophy of technology must provide an account of both the primary and the secondary instrumentalization. The existentialist tradition focused exclusively on the primary instrumentalization. Its reflections on what Peter-Paul Verbeek has called the "transcendental" preconditions of technology form the basis of a critique of modernity (Verbeek 2005). According to this critique the "essence" of technology is the orientation toward control and domination. Modern societies submit everything to technical action and so deny the intrinsic potentialities and value of all that is.

This excessively negative approach overlooks the secondary instrumentalization that complements the initial functionalization to which objects are submitted as they enter the technical field. The world is still meaningful even in the age of technology, although the meanings have certainly changed and become more fluid. The secondary instrumentalization is studied by social scientists and historians who focus precisely on what philosophers overlook: the concrete social forces at work in the design process. But without a theory of the intrinsic structure of the technical, they lack a normative perspective on the appropriate limits of technology.

These limits must be respected because the primary instrumentalization is in fact incompatible with many aspects of human life and nature. Objects introduced into technical networks bear the mark of the functionalization to which they have been submitted. Not everything of value can survive that transformation. Hence we reject the idea that more or less technically efficient means can best accomplish such things as forming friendships and enjoying Christmas dinner. The decontextualizations and reductions characteristic of the primary instrumentalization have at most a subordinate place in the background of close human relations and festive occasions.

Recontextualizing Strategies

Premodern and modern societies attach different relative weights to systematization and mediation. In premodern societies, as Latour notes, technical networks are relatively short and their nodes loosely coupled (Latour 1993). However, very elaborate valuative mediations control every aspect of technical life; here technique is inseparable from what we moderns identify as art and religion. Thus tribal weapons and huts may share a common symbolism, but they are not systematically related by technical specifications of great precision as are modern technologies. As a result, premodern societies have a limited spatial reach—their networks are confined to local regions—but they conquer time in the sense that they can be reproduced successfully over thousands of years.

Modern societies emphasize systematization and build long networks through tightly coupling links over huge distances between very different types of things and people. This requires that the artifact be stripped of most valuative mediations. The resulting overemphasis on the primary instrumentalization and systematization makes possible both large-scale hierarchical organization and technical disciplines. But despite—or perhaps because of—the power over human beings and nature they achieve, modern societies have so little control of time that it is uncertain if they will survive through the new century.

Not so long ago it was fashionable for social critics to condemn technology as such. That attitude lingers and inspires a certain haughty disdain for technology among intellectuals who nevertheless employ it constantly in their daily lives. Increasingly, however, social criticism has turned to the study and advocacy of possible reconfigurations and transformations of technology to accommodate it to actors excluded from the original design networks. This approach emerged first in the environmental movement, which was successful in modifying the design of technologies through regulation and litigation. Today it continues in proposals for transforming biotechnologies and computing.

Constructive criticism of technology takes aim at the deficiencies in the secondary instrumentalization because it is here that design receives its specific bias. This is particularly clear under capitalism, where successful business strategies often involve breaking free of various constraints on the pursuit of profits. The favored technical recontextualizations tend to ignore the values and interests of many of those involved, whether they be workers, consumers, or the community hosting production facilities. For example, in the case of logging it has been difficult until recently to convince corporations to pay attention to the health of forests and the beauty of nature. These are goods that may appeal to local communities, to sportsmen and women and environmentalists, but these actors are usually not invited to participate in the design of logging projects.

An alternative modernity worthy of the name would recover the mediating power of ethics and aesthetics. This would be accomplished not by a return to blind traditionalism but through the democratization of technically mediated institutions. Power would devolve to the members of technical networks rather than concentrating at the top of administrative hierarchies. As more actors gained access to the design process, a wider range of valuative considerations would inform technical choices. These formal changes would result in new technical designs and new ways of achieving the efficiencies that characterize modern technological activity. Whether such a society is possible or not divides critics of technology.

Real-world controversies involving technology often turn on the supposed opposition of current standards of technical efficiency and values. But this is a false opposition, as I have argued in chapter 2; current technical methods or standards were once discursively formulated as values and at some time in the past translated into the technical codes we take for granted today. This point is quite important for answering the usual so-called practical objections to arguments for social and technological reform. It seems as though the best way to do the job is compromised by attention to extraneous matters such as health or natural beauty. But the division between what appears as a condition of technical efficiency and what appears as a value external to the technical process is relative to past social and political decisions biased by unequal power. All technologies incorporate the results of such decisions and thus favor one or another actor's values or in the best of cases combine the values of several actors in clever combinations that achieve multiple goals.

This latter strategy involves what Gilbert Simondon calls "concretization," the multiplication of the functions served by the structure of a technology.[4] His illustrations of this concept are politically neutral innovations such as the air-cooled engine, which combines cooling with containment, two functions in one elegant and efficient structure. I have modified his approach to take into account what we have learned from constructivism about the social forces behind technical functions. For example, as we saw in chapter 1, the inflatable tire enabled an inherently more stable but slower bicycle design to overcome its disadvantage in bicycle racing while retaining the stability that made it attractive for transportation (Pinch and Bijker 1989, 44–46). The trade-off between speed and stability dissolved as this innovation reconciled two different social groups: young men interested in racing and ordinary riders engaged in everyday activities.

The concept of concretization explains how wider or neglected contexts can be brought to bear on technological design without loss of efficiency. A refrigerator equipped to use an ozone-safe refrigerant achieves environmental goals with the same structures that keep the milk cold. Bram Bos and his collaborators argue that industrial animal husbandry can be reorganized in ways that respect the needs of animals by employing their spontaneous behaviors in a properly reconfigured environment to protect their health and hence the efficiency of the operation (2003). What is true of devices and animals is even more true of human beings enrolled in technical networks. Their full capacities can be employed productively in a technical system designed to respect the human body and take advantage of intelligence and skill.

Technology and Democracy

Critical theory of technology is a political theory of modernity with a normative dimension. It belongs to a tradition extending from Marx to Foucault and the Frankfurt School. In this tradition, progress is analyzed as a contradictory process. Advances in the formal recognition of human rights take center stage while the insidious centralization of ever-more-powerful public institutions and private organizations imposes an authoritarian social order.

Marx attributed this pattern to the capitalist rationalization of production. Today it marks many institutions besides the factory and every modern political system, including so-called socialist systems. This pattern arose to maintain command of a disempowered and deskilled labor force, but it prevails everywhere masses are organized, whether it be Foucault's prisons or Marcuse's one-dimensional society. Technological design and development are shaped by this pattern to serve as the material base of a distinctive social order. Marcuse called this the "project" at the basis of "technological rationality." Releasing technology from this project is a democratic political task.

In accordance with this general approach, critical theory regards technologies as an environment rather than as a collection of tools. We live today with and even within the technologies that organize our way of life. Along with the constant pressures to build centers of power, many other values and meanings are inscribed in technological design. Together all these influences form a world. A hermeneutics of technology must make explicit the meanings implicit in the devices we use and the rituals they script. Social histories of technologies such as the bicycle and artificial lighting as well as studies of consumption and product design have made important contributions to this type of analysis.[5] Critical theory of technology attempts to build a methodological approach on the lessons of these accounts.

As a world, technologies shape their inhabitants. In this respect, they are comparable with laws and customs that, in antiquity, were conceived as quasi-parental influences on the citizens, raising them to conform to a desired human type. Laws, customs, and technologies can be said to both shape and represent those who live under their sway through privileging certain dimensions of their human nature. Laws of property represent the interest in ownership and control. Customs such as marriage represent the interest of the family in generational continuity and mutual support. Similarly, the automobile represents its users insofar as they are interested in mobility. Interests such as these constitute the version of human nature sanctioned by society.

Technological representation becomes salient when individuals find that important aspects of their humanity are not well served by the technical environment. Then controversies arise, as in the case of laws or customs considered unjust or outmoded. These controversies aim to alter technical designs to ensure better representation of more aspects of the humanity of users and, in some cases, victims of technology. Struggles over technology thus resemble political struggles in important respects. And in fact in the contemporary world, struggles over technology are often the most important political struggles.

Yet because the foundations of our political philosophies and constitutions were elaborated in the seventeenth and eighteenth centuries, there is still a tendency to distinguish sharply between politics and technology, the one supposedly based on rights, the other on knowledge. Much political theory argues that consensus can be achieved through the democratic exercise of those rights. In reality, political consensus is largely shaped by the available technological form of life rather than rational deliberation.

But today most technological choices are privately made and are protected from public involvement by property rights and technocratic ideology. What can be done to reverse the tide? The democratization of technology requires in the first instance the spread of knowledge, but by itself that is not enough to make a difference. In addition, the range of interests represented by those in control of technology must be enlarged so as to make it more difficult to offload externalities from technical action onto disempowered groups. Only a democratically constituted alliance of actors, embracing all those affected, is sufficiently exposed to the consequences its own actions to resist harmful projects and designs at the outset. Such broadly constituted technical alliances would take into account destructive effects of technology on the natural environment as well as on human beings. Democratic movements in the technical sphere aim to constitute such alliances. This requires a second, more narrowly political condition of technical democratization.

Democratic movements of all sorts contend with a hostile environment dominated by media manipulation. They must overcome the failure of the media to place new issues on the public agenda. Changing that agenda has been the most important achievement of the women's and environmental movements. Now the time has come for a similar change in the place of technology in public life.

Critical theory of technology projects a future in which the politics of technology is recognized as a normal aspect of public life. As with all earlier democratic movements, democracy engenders democracy: technical publics, like every earlier disempowered group, can learn from the exercise of agency how to understand their interests and constrain public institutions to serve them. The means of democratic expression on technical issues are already foreshadowed in many current practices such as hearings, citizen juries, technical controversies, protests, boycotts and legal challenges, hacking and other creative appropriations of technologies, and of course such familiar methods as elections and government regulation (Callon et al. 2009).

In a technical democracy, technical work would take on a different character. Design would be consciously oriented toward politically legitimated human values rather than subject to the whims of profit-making organizations and military bureaucracies. These values would be installed in the technical disciplines themselves, much as the value of healing presides over biological knowledge of the human body in medicine.

Classical Critical Theory was above all dedicated to interpreting the world in the light of its potentialities. Those potentialities are identified through serious study of what is and especially through the study of protests and resistances that point beyond the current horizon. Empirical research can thus be more than a mere gathering of facts and can inform an argument with the times. Philosophy of technology can join together the two extremes—potentiality and actuality, norms and facts—in a way no other discipline can rival. It must challenge the disciplinary prejudices that confine research and study and open perspectives on the future.

Conclusion

Every technology points on one side to an operator and on the other to an object. Where both operator and object are human beings, technical action is an exercise of power. Where, further, society is organized around technology, technological power is the principal form of power in the society. It is realized through designs that narrow the range of interests and concerns that can be represented by the normal functioning of the technology and the institutions that depend on it. This narrowing distorts the structure of experience and causes human suffering and damage to the natural environment.

The exercise of technical power evokes resistances of a new type immanent to the one-dimensional technical system. Those excluded from the design process eventually suffer the undesirable consequences of technologies and protest. Opening up technology to a wider range of interests and concerns would lead to its redesign for greater compatibility with human and natural limits and powers. A democratic transformation from below could shorten the feedback loops from damaged human lives and nature and guide a radical reform of the technical sphere.

An adequate understanding of the substance of our common life can no longer ignore the politics of technology. How we live is largely shaped by how we configure and design cities, transportation systems, communication media, and agricultural and industrial production. We are making more and more choices about health and knowledge in designing the technologies on which medicine, research, and education rely. Furthermore, the kinds of things it seems plausible to propose as advances or alternatives are to a great extent conditioned by the failures of the existing technologies and the possibilities they suggest. The once-controversial claim that technology is political now seems obvious.

5 From Information to Communication: The French Experience with Videotex

Information or Communication?

Notions like "postindustrial society" and the "Information Age" are forecasts—social science fictions—of a social order based on knowledge (Bell 1973). The old world of coal, steel, and railroads will evaporate in a cloud of industrial smoke as a new one based on communications and computers is born. The popularizers of this vision put a cheerful spin on many of the same trends deplored by dystopian critique, such as higher levels of organization and integration of the economy and the growing importance of expertise.

Computers play a special role in these forecasts because the management of social institutions and individual lives depends more and more on swift access to data. Not only can computers store and process data, but they can also be networked to distribute it. In the postindustrial future, computer mediated communication (CMC) will penetrate every aspect of daily life and serve the rising demand for information.

At the end of the 1960s, these predictions were taken up by political and business leaders with the power to change the world. One learns a great deal about a vision from attempts to realize it. When, as in this case, the results stray far from expectations, the theories that inspired the original forecast are called into question. This chapter explores the gap between theory and practice in a particularly important case of mass computerization: the introduction of videotex in France.

Videotex is a type of software designed for the delivery of data on computer networks. Videotex systems work as online libraries that store "pages" of information in the memory of a host computer accessible to users equipped with a terminal and modem. Today the Internet performs this function, but videotex originally pioneered a similar system. This, then, was the first major technological concretization of the notion of a post-industrial society.

The theory of the Information Age promised an emerging videotex marketplace. Experience with videotex, in turn, tested some of that theory's major assumptions in practice. Early predictions had everyone linked to videotex services well before the 1990s. By the end of the 1970s, telecommunications ministries and corporations were prepared to meet this confidently predicted future with new interactive systems. Experimental systems were launched to test different configurations of the technology. But most of these experiments were dismal failures.

This outcome may have been due in part to antitrust rulings that prevented giant telephone and computer companies from merging their complementary technologies in large-scale public systems. The Federal Communications Commission's failure to set a standard for terminals aggravated the situation. Lacking the resources and know-how of the big companies, their efforts uncoordinated by government, it is not surprising that smaller entertainment and publishing firms were unable to make a success of commercial videotex (Branscomb 1988).

Disappointing results in the United States were confirmed by all foreign experiments with videotex with the exception of the French Teletel system. The British Prestel introduced videotex three years before the French came on the scene. Ironically, the French plunged into videotex on a grand scale in part out of fear of lagging behind Britain.

Prestel had the advantage of state support, which no American system could boast. But it also had a corresponding disadvantage: overcentralization. At first, information suppliers could not connect remote hosts to the system, which severely limited growth in services. What is more, Prestel relied on users to buy a decoder for their television sets, an expensive piece of hardware that placed videotex in competition with television programming. The subscriber base grew with pathetic slowness, rising to only seventy-six thousand in the first five years (Charon 1987, 103–106; Mayntz and Schneider 1988, 278).

Meanwhile, the successful applications of CMC were all organized by and for private businesses, universities, or computer hobbyists. The general public still had little or no access to these networks and no need to use such specialized online services as bibliographic searches and software banks. Thus after a brief spurt of postindustrial enthusiasm for videotex, CMC was regarded as suitable primarily for work, not for pleasure; it was expected to serve professional needs rather than leisure or consumption (Ettema 1989).

As I will explain later, the Teletel story is quite different. Between 1981, the date of the first tests of the French system, and the end of the decade, Teletel became by far the largest public videotex system in the world, with thousands of services, millions of users, and hundreds of millions of dollars in revenues. Until it was eclipsed by the Internet, Teletel was the brightest spot in the otherwise unimpressive commercial videotex picture. (Teletel still exists, but the Internet has replaced it for many purposes.)

This outcome is puzzling. Could it be that the French are different from everyone else? That rather silly explanation became less plausible as Compu-Serve and the Sears/IBM Prodigy system grew to a million subscribers in the early 1980s. The sheer size of these earlier systems confirmed the existence of a home videotex market, but at first only the French knew how to profit from it. How, then, can we account for the astonishing success of Teletel, and what are its implications for the Information Age theory that inspired its creation?

Teletel is particularly interesting because it employed no technology not readily available in all those other countries where videotex was tried and failed. Its success can be explained only by identifying the *social inventions* that aroused widespread public interest in CMC. A close look at those inventions shows the limitations not only of prior experiments with videotex but also of the theory of the Information Age (Feenberg 1991, chap. 5).

The Emergence of a New Medium

While Teletel embodies generally valid discoveries about domestic CMC, it is also peculiarly French. Much that is unique about it stems from the confluence of three factors: (1) a specifically French politics of modernization; (2) the bureaucracy's voluntaristic ideology of national public service; and (3) a strong oppositional political culture. Each of these factors contributed to a result no single group in French society would willingly have served in the beginning. Together they opened the space of social experimentation that Teletel made technically possible.

Modernization

The concept of modernity is a live issue in France in a way that is difficult to imagine in the United States. Americans experience modernity as a birthright; America does not *strive* for modernity, it *defines* modernity, or at least so it believes. For that reason, the United States does not treat its own modernization as a political issue but relies on the creative chaos of the market.

France, on the other hand, has a long tradition of theoretical and political concern with modernity as such. In the shadow of England at first and later of Germany and the United States, France has struggled to adapt itself to a modern world it has always experienced to some extent as an external challenge. The extraordinary backwardness of the French telephone system was a symbol of this general conservatism, and so its rapid modernization under President Giscard d'Estaing signified the will to meet the challenge. This is the spirit of the famous *Nora-Minc Report*, which Giscard commissioned from two top civil servants to define the means and goals of a concerted policy of modernization for French society in the last years of the century (Nora and Minc 1978).

Nora and Minc called for a technological offensive in "telematics," the term they coined to describe the marriage of computers and communications. The telematic revolution, they argued, will change the nature of modern societies as radically as did the Industrial Revolution. But, they added, "'Telematics,' unlike electricity, does not carry an inert current, but rather information, that is to say, power." "Mastering the network is therefore an essential goal. Its framework must therefore be conceived in the spirit of public service" (1978, 11, 67). In sum, just as war is too important to be left to the generals, so postindustrial development is too important to be left to businesspeople and must become a political affair.

Nora and Minc paid particular attention to the need to win public acceptance of the telematic revolution and to achieve success in the new international division of labor through targeting emerging telematic markets (1978, 41–42). They argued that a national videotex service could play a central role in achieving these objectives. This service would sensitize the still-backward French public to the wonders of the Information Age while creating a huge protected market for computer terminals. Leveraging the internal market, France would eventually become a leading exporter of terminals and so benefit from the expected restructuring of the international economy instead of falling further behind (1978, 94–95). These ideas lay at the origin of the Teletel project, which, as a peculiar mix of propaganda and industrial policy, had a distinctly statist flavor from the very beginning.

Voluntarism

So conceived, the project fell naturally into the hands of the civil service. This choice, which seems strange to Americans contemptuous of bureaucratic ineptness, makes perfect sense in France, where business has an even more negative image than government.

When it is the bureaucracy rather than the corporation that spearheads modernization, the esprit de corps of the civil service leaves its mark on the outcome. In France this is not such a bad thing. French bureaucrats define the nation in terms of the uniform provision of services such as mail, phone, roads, schools, and so on. Delivering these services is a moral mission predicated on the "republican" ideal of egalitarianism. The French call this bureaucratic approach "voluntaristic" because, for better or worse, it ignores local situations and economic constraints to serve a universal public interest.

One must keep this voluntaristic sense of mission in mind to understand how the government-owned French telephone company, charged with developing Teletel, could have conceived and implemented a national videotex service without any guarantee of profitable operation. In fact, Teletel was less a money-making scheme than a link in the chain of national identity. As such, it was intended to reach every French household as part of the infrastructure of national unity, just like the telephone and the mails (Nora and Minc 1978, 82).

To achieve this result, the telephone company proceeded to distribute millions of free terminals, called "Minitels." Although early advertising was mainly directed at prosperous neighborhoods, anyone could request a Minitel. Eventually all phone subscribers were to be equipped. France would leapfrog out of its position as the industrial country with the most backward telephone system right into the technology of the next century.

An American telephone company would certainly have charged for such an elaborate upgrade of the users' equipment. Even the French government was a bit worried about justifying this unprecedented bounty. The excuse it came up with was the creation of a national electronic phone directory, accessible only by Minitel, but in fact the main point of the exercise was simply to get a huge number of terminals out the door as quickly as possible (Marchand 1987, 32–34). Free distribution of terminals preceded the development of a market in services, which it was supposed to bring about. Just as roadside businesses follow highways, so telematic businesses were expected to follow the distribution of Minitels.

The first four thousand Minitels were delivered in 1981 (Marchand 1987, 37); ten years later over five million had been distributed. The speed and scale of this process are clues to the economics of the great telematic adventure. The telephone company's ambitious modernization program had made it the largest single customer for French industry in the 1970s. The daring telematic plan was designed to take up the slack in telephone production that was sure to follow the saturation of that market, thereby avoiding the collapse of a major industrial sector.

Opposition

As originally conceived, Teletel was designed to bring France into the Information Age by providing a wide variety of services. But is more information what every household needs (Iwaasa 1985, 49)? And who is qualified to offer information services in a democracy (Marchand 1987, 40ff)? These questions received a variety of conflicting answers in the early years of French videotex.

Modernization through national service defines the program of a highly centralized and controlling state. To make matters worse, the Teletel project was initiated by a conservative government. This combination at first inspired widespread distrust and awakened the well-known fractiousness of important sectors of opinion. The familiar pattern of central control and popular resistance was repeated once again with Teletel, a program that was "parachuted" onto an unsuspecting public and soon transformed by it in ways its makers had never imagined.

The press led the struggle against government control of videotex. When the head of the French telephone company announced the advent of the paperless society (in Dallas, of all places), publishers reacted negatively out of fear of losing advertising revenues and independence. The dystopian implications of a computer-dominated society did not pass unnoticed. One irate publisher wrote, "He who grasps the wire is powerful. He who grasps the wire and the screen is very powerful. He who will someday grasp the wire, the screen, and the computer will possess the power of God the Father Himself" (quoted in Marchand 1987, 42).

The press triumphed with the arrival of the socialist government in 1981. To prevent political interference with online content, the telephone company itself was allowed to offer only its electronic version of the telephone directory. Meanwhile, the doors to Teletel were opened wide by the standards of the day: anyone with a publisher's license could connect a host to the system. In 1986 even this restriction was abandoned; today anyone with a computer can hook up to the system, list a phone number in the directory, and receive a share of the revenues the service generates for the phone company.

Because small host computers are fairly inexpensive and knowledge of videotex no more common in large than in small companies, these decisions had at first a highly decentralizing effect. Teletel became a vast space of disorganized experimentation, a "free market" in online services more nearly approximating the liberal ideal than most communication markets in contemporary capitalist societies.

This example of the success of the market has broad implications, but not quite so broad as the advocates of deregulation imagine. The fact that markets sometimes mediate popular demands for technical change does not make them a universal panacea. All too often markets are manipulated by large corporations to sell well-established technologies and stifle the demands that existing products cannot meet or rechannel those demands into domains where basic technical change need not occur. Nevertheless, consumers do reopen the design process through the market. This is certainly a reason to view markets as ambivalent institutions with a potentially dynamic role to play in the development of new technology.

Communication

Surprisingly, although phone subscribers were now equipped for the Information Age, they made relatively little use of the wealth of data available on Teletel. They consulted the electronic directory regularly but not much else. Then, in 1982, hackers transformed the technical support facility of an information service called "Gretel" into a messaging system (Bruhat 1984, 54–55). After putting up a feeble (perhaps feigned) resistance, the operators of this service institutionalized the hackers' invention and made a fortune. Other services quickly followed with names like "Désiropolis,"

"La Voix du Parano," "SM," "Sextel." "Pink" messaging became famous for spicy pseudononymous conversations in which users sought like-minded acquaintances for conversation or encounters.

Once messaging took off on a national scale, small telematic firms reworked Teletel into a communication medium. They designed programs to manage large numbers of simultaneous users emitting as well as receiving information, and they invented a new type of interface. On entering these systems, users are immediately asked to choose a pseudonym and to fill out a brief CV (curriculum vitae, or *carte de visite*). They are then invited to survey the CVs of those currently online to identify like-minded conversational partners. The programs employ the Minitel's graphic capabilities to split the screen, assigning each of as many as a half dozen communicators a separate space for their messages. This is where the creative energies awakened by telematics went in France and not into meeting obscure technical challenges dear to the hearts of government bureaucrats such as ensuring French influence on the shape of the emerging international market in databases (Nora and Minc 1978, 72).

The original plans for Teletel had not quite excluded human communication, but its importance relative to the dissemination of data, online transactions, and even video games were certainly underestimated (Marchand 1987, 136). Messaging is hardly mentioned in early official documents on telematics (e.g., Pigeat et al. 1979). The first experiment with Teletel at Vélizy revealed an unexpected enthusiasm for communication. Originally conceived as a feedback mechanism linking users to the Vélizy project team, the messaging system was soon transformed into a general space for free discussion (Charon and Cherky 1983, 81–92; Marchand 1987, 72). Even after this experience, no one imagined that human communication would play a major role in a mature system. But that is precisely what happened.

In the summer of 1985, the volume of traffic on Transpac, the French packet-switching network, exceeded its capacities, and the system crashed. The proud champion of French high tech was brought to its knees as banks and government agencies were bumped offline by hundreds of thousands of users skipping from one messaging service to another in search of amusement. This was the ultimate demonstration of the new telematic dispensation (Marchand 1987, 132–134). Although only a minority of users was involved, by 1987 40 percent of the hours of domestic traffic were spent on messaging (Chabrol and Perin 1989, 7).

"Pink" messaging may seem a trivial result of a generation of speculation on the Information Age, but the case can be made for a more positive
evaluation. Most importantly, the success of messaging changed the generally received *imaginaire* of telematics, away from information toward
communication.[1] This in turn encouraged—and paid for—a wide variety
of experiments in domains such as education, health, and news (March-
and 1987; Bidou et al. 1988). Here are some examples:

• television programs offered services through which viewers obtained
supplementary information or exchanged opinions, adding an interactive
element to the one-way broadcast;

• politicians engaged in dialog with constituents, and political movements
opened messaging services to communicate with their members;

• educational experiments brought students and teachers together for electronic classes and tutoring, for example, at a Paris medical school;

• a psychological service offered an opportunity to discuss personal problems anonymously and seek advice;

• the messaging service of the newspaper *Libération* coordinated a national
student strike in 1986. Perhaps the most interesting of the messaging experiments, the service offered information about issues and actions, online
discussion groups, hourly news updates, and a game mocking the minister
of education. It quickly received three thousand messages from all over the
country (Marchand 1987, 155–158). This must be one of the first if not the
first application of electronic networking to public protest.

These applications revealed the unsuspected potential of CMC for creating surprising new forms of sociability. Rather than imitating the telephone or writing, they play on the unique capacity of telematics to mediate
highly personal and often anonymous communication. These experiments prefigured a very different organization of public and private life in
advanced societies, the full extent of which begins to be visible with Web
2.0 (Feenberg, 1989a: 271–275; Jouet and Flichy, 1991.)

The System

Although no one planned all its elements in advance, eventually a coherent system emerged from the play of these various forces. Composed of
rather ordinary elements, it formed a unique whole that finally broke the

barriers to general public acceptance of CMC. The system was characterized by five basic principles:

1. Scale   Only a government or a giant corporation has the means to ini-
tiate an experiment such as Teletel on a large enough scale to ensure a fair
test of the system. Smaller pilot projects all foundered on a chicken-and-
egg dilemma: to build a market in services one needs users, but users cannot be attracted without a market in services. The solution, demonstrated
in France, was to make a huge initial investment in transmission facilities
and terminals in order to attract enough users at an early stage to justify
the existence of a critical mass of services.[2]

2. Gratuity   Perhaps the single most revolutionary feature of the system
was the free distribution of terminals. The packet-switching network
and the terminals were treated as a single whole, in contradistinction
to every other national computer network. Gratuity dictated wise decisions about terminal quality. The emphasis was on durability and simple
graphics and interface. It also ensured service providers a large base to
work from very early on, long before the public would have perceived the
interest of the unfamiliar system and invested in a costly terminal or
subscription.

3. Standardization   The monopoly position of the French telephone company and the free distribution of Minitel terminals ensured uniformity in
several vital areas. Equipment and sign-on procedures are standardized,
and a simple navigational interface resembling a Web browser is built into
the terminal keyboard. Most service is offered from a single national phone
number at a single price, independent of location. The phone company
employs its billing system to collect all charges, sharing the income with
service providers.

4. Liberalism   The decision to make it easy to hook up host computers to
the packet-switching network must have gone against the telephone company's ingrained habit of controlling every aspect of its technical system.
However, once this decision was made, it opened the doors to a remarkable flowering of social creativity. Although the Minitel was designed primarily for information retrieval, it can be used for many other purposes.
The success of the system owes a great deal to the mating of a free market
in services with the flexible terminal.

5. Identity The system acquired a public image through its identification with a project of modernization and through the massive distribution of distinctive terminals. A unique telematic image was also shaped by the special phone directory, the graphic style associated with Teletel's alpha-mosaic standard, the adoption of videotex screen management instead of scrolling displays, and the social phenomenon of the "pink" messaging.

The Conflict of Codes

This interpretation of Teletel contradicts the deterministic assumptions about the social impact of computers that inspired Nora, Minc, and many other theorists of postindustrialism. The logic of technology simply did not dictate a neat solution to the problem of modernization; instead, a very messy process of conflict, negotiation, and innovation produced a socially contingent result. What were these social factors, and how did they influence the development of CMC in France?

Social Constructivism

Teletel's evolution confirms the social constructivist approach introduced in previous chapters. Unlike determinism, social constructivism does not explain the success of an artifact by its technical characteristics. According to the "principle of symmetry," there are always alternatives that might have been developed in the place of the successful one. What singles out an artifact is not some intrinsic property such as "efficiency" or "effectiveness" but its relationship to the social environment.

As we have seen in the case of videotex, that relationship is negotiated among inventors, civil servants, businesspeople, consumers, and many other social groups in a process that ultimately defines a specific product adapted to a specific mix of social demands. This process ends in "closure"; it produces a stable "black box," an artifact that can be treated as a finished whole. Before a new technology achieves closure, its social character is evident, but once it is well established, its development appears purely technical, even inevitable to a naïve backward glance. Typically, later observers forget the original ambiguity of the situation in which the "black box" was first closed (Latour 1987, 2–15).

This approach has several implications for videotex:

• First, the design of a system like Teletel is not determined by a universal criterion of efficiency but by a social process that judges technical alternatives according to a variety of criteria.

• Second, that social process is not about the application of a predefined videotex technology but concerns the very definition of videotex and the nature of the problems to which it is addressed.

• Third, competing definitions reflect conflicting social visions of modern society concretized in different technical choices.

• Fourth, new social groups and categories emerge around the appropriation of new technology or resistance to its impacts, leading to design changes.

These four points indicate the need for a revolution in the study of technology. The first point widens the range of social conflict to include technical issues that, typically, have been treated as the object of a purely "rational" consensus. The next two points imply that meanings enter history as effective forces not only through cultural production and political action but also in the technical sphere. Understanding the social perception or definition of a technology requires a hermeneutic of technical objects. The last point introduces the co-construction of society and technology.

Technologies are meaningful objects. From our everyday commonsense standpoint, two types of meanings attach to them. In the first place, they have a function, and for most purposes their meaning reflects that function. However, we also recognize a penumbra of "connotations" that associates technical objects with other aspects of social life independent of function (Baudrillard 1968, 16–17). Thus automobiles are means of transportation, but they also signify the owner as more or less respectable, wealthy, sexy, and so forth.

In the case of well-established technologies, the distinction between function and connotation is usually clear. There is a tendency to project this clarity back into the past and to imagine that the technical function preceded the object and called it into being. The social constructivist program argues, on the contrary, that technical functions are not pregiven but are discovered in the course of the development and use of the object. Gradually certain functions are locked in by the evolution of the social and technical environment. For example, the transportation functions of the automobile have been institutionalized in low-density urban designs that create the demand automobiles satisfy. Closure thus depends in part on building tight connections in a larger technical network.

In the case of new technologies, there is often no clear definition of function at first. As a result, there is no clear distinction between different types of meanings associated with the technology. Recall Pinch and Bijker's example of the bicycle discussed in chapter 1. Connotations of one design may be functions viewed from the angle of the other. These ambiguities are not merely conceptual if the device is not yet "closed" and no institutional lock-in ties it decisively to one of its several uses. Thus ambiguities in the definition of a new technology must be resolved through technical development itself. Designers, purchasers, and users all play a role in the process by which the meaning of a new technology is finally fixed.[3]

Technological closure is eventually consolidated in a technical code. Technical codes define the object in strictly technical terms in accordance with the social meanings it has acquired. For bicycles, this was achieved in the 1890s. A bicycle safe for transportation could be produced only in conformity with a code dictating a seat positioned well behind a small front wheel. When consumers encountered a bicycle produced according to this code, they immediately recognized it for what it was: a "safety" in the terminology of the day. That definition in turn connoted women and older riders, trips to the grocery store, and so on and negated associations with young sportsmen out for a thrill.

Technical codes are interpreted with the same hermeneutic procedures used to interpret texts, works of art, and social actions (Ricoeur 1979). But the task gets complicated when codes become the stakes in significant social disputes. Then ideological visions are sedimented in design. Hence the "isomorphism, the formal congruence between the technical logics of the apparatus and the social logics within which it is diffused" (Bidou et al. 1988, 18). These patterns of congruence explain the impact of the larger socio-cultural environment on the mechanisms of closure (Pinch and Bijker 1989, 46). Videotex is a striking case in point. In what follows I will trace the pattern from the macrolevel of worldviews down to the details of technical design.

A Technocratic Utopia

The issue in this case is the very nature of a postindustrial society. The Information Age was originally conceived as a scientized society, a vision that legitimated the technocratic ambitions of governments and corporations. The rationalistic assumptions about human nature and society that underlie this fantasy have been familiar for a century or more as a kind of positivist utopia.

Its principal traits are familiar. Scientific-technical thinking becomes the logic of the whole social system. Politics is merely a generalization of the consensual mechanisms of research and development. Individuals are integrated into the social order not through repression but through prosperity. Their well-being is achieved through technical mastery of the personal and natural environment. Power, freedom, and happiness are thus all based on knowledge.

This global vision supports the generalization of the codes and practices associated with engineering and management. One need not share an explicit utopian faith to believe that the professional approaches of these disciplines are useful outside the contexts in which they are customarily applied. The spread of ideas of social engineering based on systems analysis, rational choice theory, risk/benefit analysis, and so on testifies to this advance in the rationalization of society. Similar assumptions influenced the sponsors of Teletel, not surprisingly given the cult of engineering in the French bureaucracy.

At the microlevel, these assumptions are at work in the traditional computer interface, with its neat menu hierarchies consisting of one-word descriptors of "options." A logical space consisting of such alternatives correlates with an individual "user" engaged in a personal strategy of optimization. Projected onto society as a whole in the form of a public information service, this approach implies a certain world.

In that world, "freedom" is the more or less informed choice among preselected options defined by a universal instance such as a technocratic authority. That instance claims to be a neutral medium, and its power is legitimated precisely by its transparency: the data are accurate and logically classified. But it does not cease to be a power for that matter.

Individuals are caught up in just such a system as this in their interactions with corporate, government, medical, and scholastic institutions. Videotex streamlines this technocratic universe. In fact some of the most successful utilitarian services on Teletel offer information on bureaucratic rules, career planning, or examination results. These services play on the "anxiety effect" of life in a rational society: individuality as a problem in personal self-management (Bidou et al. 1988, 71). But the role of anxiety reveals the darker side of this utopia. The system appears to embody a higher level of social rationality, but it is a nightmare of confusing complexity and arbitrariness to those whose lives it shapes. This is the "Crystal Palace" so feared and hated in Dostoyevsky's "underground" or Godard's *Alphaville,* where the computer's benign rule is the ultimate dehumanizing oppression.

The Spectral Subject

Teletel was caught up in a dispute over which sort of postindustrial experience would be projected technologically through domestic computing. As we have seen, the definition of interactivity in terms of a rationalistic technical code encountered immediate resistance from users who ignored the informational potential of the system and instead employed it for anonymous human communication.

This unexpected application revealed another whole dimension of everyday experience in postindustrial societies masked by the positivist utopia. As the gap between individual person and social role widens, and individuals are caught up in the "mass," social life is increasingly reorganized around impersonal interactions. The individual slips easily between roles and identifies fully with none of them, falls in and out of various masses daily, and belongs wholly to no community. The solitude of the "lonely crowd" consists in a multitude of trivial and ambiguous encounters. The simplified codes of interaction in the "system" offer few possibilities of personal self-expression or attachment to others. Anonymity plays a central role in this new social experience and gives rise to fantasies of sex and violence that are represented in mass culture and, to a lesser extent, realized in the individuals' lives.

Just as videotex permits the individual to personalize an anonymous query to a career planning agency or a government bureaucracy, so the hitherto inarticulate relationship to erotic texts can now achieve personality, even reciprocity, thanks to the Minitel. The privacy of the home takes on functions previously assigned public spaces like bars and clubs, but with an important twist: the blank screen not only links the interlocutors but also shields their identities.

As with newspaper "personals," individuals have the impression that the Minitel gives them full command of all the signals they emit, unlike risky face-to-face encounters where control is uncertain at best. Enhanced control through written self-presentation makes elaborate identity games possible. "Instead of identity having the status of an initial given (with which the communication usually begins), it becomes a stake, a product of the communication" (Baltz 1984, 185).

The experience of pseudononymous communication calls to mind Erving Goffman's double definition of the self as an "image" or identity and as a "sacred object" to which consideration is due: "the self as an image pieced together from the expressive implications of the full flow of events in an undertaking; and the self as a kind of player in a ritual game who copes honorably or dishonorably, diplomatically or undiplomatically, with the judgmental contingencies of the situation" (1982, 31). By increasing control of image while diminishing the risk of embarrassment, messaging alters the sociological ratio of the two dimensions of selfhood and opens up a new social space.

The relative desacralization of the subject weakens social control. It is difficult to bring group pressure to bear on someone who cannot see frowns of disapproval. CMC thus enhances the sense of personal freedom and individualism by reducing the "existential" engagement of the self in its communications. "Flaming"—the expression of uncensored emotions online—is a negative consequence of this feeling of liberation. But the altered sense of the reality of the other may also enhance the erotic charge of the communication (Bidou et al. 1988, 33).

Marc Guillaume has introduced the concept of "spectrality" to describe these new forms of interaction between individuals who are reduced to anonymity in modern social life and use that anonymity to shelter and assert their identities.

Teletechnologies, considered as a cultural sphere, respond to a massive and unconfessed desire to escape partially and momentarily both from the symbolic constraints which persist in modern society and from totalitarian functionality. To escape not in the still ritualized form of those brief periods of celebration or disorder permitted by traditional societies, but at the convenience of the subject, who pays for this freedom by a loss. He becomes a *specter* . . . in the triple sense of the term: he fades away in order to wander freely like a phantom in a symbolic order which has become transparent to him (1982: 23, my translation).

Social advance appears here not as the spread of technocratic elements throughout daily life but as the generalization of the commutative logic of the telephone system. National computer networks such as Teletel are based on the X.25 standard, which enables host computers to serve distant "clients" through the telephone lines. Although such networks can link all their hosts much as the telephone system links all subscribers, that is not what they were originally designed to do. Rather they were supposed to enable clusters of users to share time on specialized hosts. In the usual case, the users are not in communication with each other.

Teletel started out as an ordinary X.25 network in which the user is a point in a star-shaped interaction, hierarchically structured from a center, the host computer. But in the practice of the system, the user became an agent of general horizontal interconnection (Guillaume, 1986: 177ff). This shift symbolizes the emergence of "networking" as an alternative to both formal organization and traditional community. The computer system provides a particularly favorable environment in which to experiment with this new social form.

In CMC the pragmatics of personal encounter are radically simplified, in fact reduced, to the protocols of technical connection. Correspondingly, the ease of passage from one social contact to another is greatly increased, again following the logic of commutation. "Pink" messaging is merely a symptom of this transformation, punctuating a gradual process of change in society at large. To fully understand this alternative, it is once again useful to look at the technical metaphors that invade social discourse.

A whole rhetoric of liberation accompanies the generalized breakdown of the last rituals blocking the individuals in the redoubt of the sacred self. Personal life becomes an affair of network management as family and other stable structures collapse. The new postmodern individuals are described as supple, adaptable, capable of staging their personal performances on many and changing scenes from one day to the next. The network multiplies the power of its members by joining them in temporary social contracts along digital pathways of mutual confidence. The result is a postmodern "atomisation of society into flexible networks of language games" (Lyotard 1979, 34).

Teletel profoundly altered the spatio-temporal coordinates of daily life, accelerating the individuals beyond the speed of paper, which was still the maximum velocity achieved by shuffling corporate and political dinosaurs. Users achieved thereby a relative liberation: if you cannot escape the postindustrial nightmare of total administration, at least multiply the number of connections and contacts so that their point of intersection becomes a rich and juicy locus of choice. To be is to connect. Thus begins the struggle over the definition of the postindustrial age.

The Social Construction of the Minitel

The peculiar compromise that made Teletel a success was the resultant of these forces in tension. I have traced the terms of that compromise at the macrolevel of the social definition of videotex in France, but its imprint can also be identified in the technical code of the system interface.

Wiring the Bourgeois Interior

The Minitel is a sensitive index of these tensions. Those charged with designing it feared public rejection of anything resembling a computer, typewriter, or other professional apparatus and worked to fit it into the domestic environment. They carefully considered the "social factors" as well as the human factors involved in persuading millions of ordinary people to admit a terminal into their home (Feenberg 1989b, 29).

This is a design problem with a long and interesting history. Its presupposition is the separation of public and private, work and home, which begins, according to Walter Benjamin, under the July Monarchy: "For the private person, living space becomes, for the first time, antithetical to the place of work. The former is constituted by the interior; the office is its complement. The private person who squares his accounts with reality in his office demands that the interior be maintained in his illusions" (Benjamin 1978, 154).

The history of design shows these intimate illusions gradually shaped by images drawn from the public sphere through the steady invasion of private space by public activities and artifacts. Everything from gas lighting to the use of chrome in furniture begins life in the public domain and gradually penetrates the home (Schivelbusch 1988; Forty 1986, chap. 5). The telephone and the electronic media intensify the penetration by decisively shifting the boundaries between the public and the private spheres.

The final disappearance of what Benjamin calls the "bourgeois interior" awaits the generalization of interactivity. The new communications technologies promise to attenuate and perhaps even to dissolve the distinction between the domestic and the public spheres. Telework and telemarketing are expected to collapse the two worlds into one. "The home can no longer pretend to remain the place of private life, privileging noneconomic relations, autonomous with respect to the commercial world" (Marchand 1984, 184).

The Minitel is a tool for accomplishing this ultimate deterritorialization. Its modest design is a compromise on the way toward a radically different type of interior. Earlier videotex systems had employed very elaborate and expensive dedicated terminals, television adapters, or computers equipped with modems. In the United States, domestic CMC was computer based. Its spread had to await the generalization of computer ownership. Until then it was largely confined to a hobbyist subculture. No design principles for the Minitel could be learned from these hobbyists, who were not bothered by the incongruous appearance of a large piece of electronic equipment on the bedroom dresser or the dining room table. Functionally, the Minitel is not even a real computer. It is just a "dumb terminal," that is, a video screen and keyboard with minimal memory and processing capabilities and a built-in modem. Such devices had been around for decades, primarily for use by engineers to operate mainframe computers. They were generally large, expensive, and ugly. Obviously those designs would not qualify as attractive interior decoration.

The Minitel's designers broke with all these precedents and connoted it as an enhancement of the telephone rather than as a computer or a new kind of television (Giraud 1984, 9). Disguised as a "cute" telephonic device, the Minitel was a kind of Trojan horse for rationalistic technical codes.

It is small with a keyboard that can be tilted up and locked to cover the screen. At first it was equipped with an alphabetical keypad to distinguish it from a typewriter. That keypad pleased neither nontypists nor typists and was eventually replaced with a standard one; however, the overall look of the Minitel remained unbusinesslike (Marchand 1987, 64; Norman 1988, 147). Most important, it has no disks and disk drives, the on-off switch on its front is easy to find, and no intimidating and unsightly cables—just an ordinary telephone cord—protrude from its back.

The domesticated Minitel terminal adopts a telephonic rather than a computing approach to its users' presumed technical capabilities. Computer programs typically offer an immense array of options, trading off ease of use for power. Furthermore, until the success of Windows, most programs had such different interfaces that each one required a special apprenticeship. Anyone who has ever used early DOS communications software, with its opening screens for setting a dozen obscure parameters, can understand just how inappropriate it would be for general domestic use. The Minitel designers knew their customers well and offered an extremely simple connection procedure: dial up the number on the telephone, listen for the connection, press a single key.

The design of the function keys also contributed to ease of use. These were intended to operate the electronic telephone directory. At first there was some discussion of giving the keys highly specific names suited to that purpose, for example, "City," "Street," and so on. It was wisely decided instead to assign the function keys general names, such as "Guide," "Next Screen," "Back," rather than tying them to any one service (Marchand 1987, 65). As a result, the keyboard imposes a standard navigational user interface not unlike the World Wide Web, something achieved in the computing world only much later with much more elaborate equipment.

The Minitel testifies to the designers' original skepticism with regard to communication applications of the system: the function keys are defined for screen-oriented interrogation of databases, and the keypad, with its unsculptured chiclet keys, is so clumsy it defies attempts at touch typing. Here the French paid the price of relying on a telephonic model: captive Telecom suppliers ignorant of consumer electronics markets delivered a telephone-quality keypad below international standards for even the cheapest portable typewriter. Needless to say, export of such a terminal was next to impossible.

Ambivalent Networks

So designed, the Minitel is a paradoxical object. Its telephonic disguise, thought necessary to its success in the home, introduces ambiguities into the definition of telematics and invites communications applications not anticipated by the designers (Weckerlé 1987, I, 14–15). For them the Minitel would always remain a computer terminal for gathering data, but the domestic telephone, to which the Minitel is attached, is a social, not an informational medium. The official technical definition of the system thus enters into contradiction with the telephonic practices that immediately colonize it once it is installed in the home (Weckerlé 1987, I, 26).

To the extent that the Minitel did not rule out human communication altogether, as have many videotex systems, it could be subverted from its intended purpose despite its limitations. For example, although the original function keys were not designed for messaging applications, they could be incorporated into messaging programs, and users adapted to the poor keyboard by typing in an online shorthand rich in new slang and inventive abbreviations. The Minitel thus became a communication device.

The walls of Paris were soon covered with posters advertising messaging services. A whole new iconography of the reinvented Minitel replaced the sober modernism of official PTT propaganda. In these posters, the device is no longer a banal computer terminal but is associated with blatant sexual provocation. In some ads, the Minitel walks, it talks, it beckons; its keyboard, which can flap up and down, becomes a mouth, the screen becomes a face. The silence of utilitarian telematics is broken in a bizarre cacophony.

In weakening the boundaries of private and public, the Minitel opens a two-way street. In one direction, households become the scene of hitherto public activities, such as consulting train schedules or bank accounts. But in the other direction, telematics unleashes a veritable storm of private fantasy on the unsuspecting public world. The individual still demands, in Benjamin's phrase, that the "interior be maintained in his illusions." But now those illusions take on an aggressively erotic aspect and are broadcast over the network.

The technical change in the Minitel implied by this social change is invisible but essential. It was designed as a client node, linked to host computers, and was not intended for use in a universally switched system that, like the telephone network, allows direct connection of any subscriber with any other. As its image changed, the telecom responded by creating a universal electronic mail service, called "Minicom," which offered a mailbox to everyone with a Minitel. But unfortunately, a new group of bureaucrats managing the system lacked the imagination and daring of its originators. Minicom was household based. Unless one lived alone, it was impossible to engage in private exchanges on this service. Needless to say, it never enjoyed the success of email on the Internet.

Despite the revenues earned from these communications applications, the French Telecom grumbled that its system was being misused. Curiously, those who introduced the telephone a century ago fought a similar battle with users over its definition. The parallel is instructive. At first the telephone was compared with the telegraph and advertised primarily as an aid to commerce. In opposition to this "masculine" identification of the telephone, women gradually incorporated it into their daily lives as a social instrument (Fischer, 1988b). There was widespread criticism of social uses of the telephone, and an attempt was made to confine it to a business role (Fischer 1988a; Attali and Stourdzé 1977). As one telephone company official complained in 1909: "The telephone is going beyond its original design, and it is a positive fact that a large percentage of telephones in use today on a flat rental basis are used more in entertainment, diversion, social intercourse and accommodation to others than in actual cases of business or household necessity" (quoted in Fischer 1988a, 48).

In France erotic connotations clustered around these early social uses of the telephone. It was worrisome that outsiders could intrude on the home while the husband and father were away at work. "In the imagination of the French of the *Belle Epoque*, the telephone was an instrument of seduction" (Bertho 1981, 243). So concerned was the phone company for the virtue of its female operators that it replaced them at night with males, presumably proof against temptation (Bertho 1981, 242–243).

Despite these difficult beginnings, by the 1930s sociability had become an undeniable referent of the telephone in the United States. (In France the change took longer.) Thus the telephone is a technology that, like videotex, was introduced with an official definition rejected by many users. And like the telephone, the Minitel acquired new and unexpected functions as it became a privileged instrument of personal encounter. In both cases, the magic play of presence and absence, of disembodied voice or text, generates unexpected social possibilities inherent in the very nature of mediated communication.

Conclusion: From Teletel to the Internet

In its final configuration, Teletel was largely shaped by the users' preferences (Charon 1987, 100). The picture that emerges is quite different from initial expectations. What are the lessons of this outcome? The rationalistic image of postindustrial society did not survive the test of experience unchanged. Teletel is not just an information marketplace. Alongside the expected applications, users invented a new form of human communication to suit the need for social play and encounter in an impersonal, bureaucratic society. In so doing, ordinary people overrode the intentions of planners and designers and converted an informational resource into a postmodern social environment.

The meaning of videotex technology was irreversibly changed by this experience. When the Internet was opened to the public, similar user initiatives resulted in the proliferation of new social forms on a system originally designed for time sharing on mainframe computers.

If the Internet was ultimately more successful, this is due to its unusual technical design. Unlike the X.25 networks created by national telecoms, the Internet enabled each computer connected to the system to manage its own data. The system spread wherever personal computers were in use without regard for local standards of the sort imposed by the French and other national telecoms. The result has been the emergence of a global communication system supporting an unprecedented variety of activities.

But beyond these particulars, a larger picture looms. In every case, the human dimension of communication technology emerges only gradually in opposition to the cultural assumptions of those who originate it and first signify it publicly through rationalistic codes. This process reveals the limits of postindustrial ideology.

6 Technology in a Global World

Introduction

Japan has always been the test case for the universality of Western culture. The Japanese were the first non-Western people to modernize successfully. They built a powerful economy based on Western science and technology. Yet their society remains significantly different from the Western models it imitates. These differences are not merely superficial vestiges of a dying tradition but show up in the very structure of Japanese science and technology (Traweek 1988). Is Japan different enough to qualify as an "alternative modernity"? Does it refute or confirm the claims of universalism? These are the questions Japan raises for us today. An early response to these questions comes from Japan itself. In the 1930s, the founder of modern Japanese philosophy, Kitaro Nishida, proposed an innovative theory of multicultural modernity. In this chapter, I will consider the Japanese case and introduce Nishida's remarkable theory, one of the first attempts to grasp the philosophical implications of globalization. In conclusion I will show how the Japanese response to technological modernization foreshadows later problems in the West.

Two Types of Technological Development

Japan cut off nearly all relations with the rest of the world from the early seventeenth century until the country was forcibly opened to trade by American warships in 1854. Thereafter, Japan modernized with incredible rapidity. The modernization process touched every aspect of life, including shopping.

The department store was introduced into Japan during the late Meiji era (1868–1912) by the Mitsui family. Its store, Mitsukoshi, was successful and expanded until it was as large as the Western department stores it imitated (Seidensticker 1983).

However, in one respect the Japanese store was quite different from its models: Mitsukoshi had tatami mat floors. This made for some unique problems. Japanese consumers did not usually remove their shoes to enter the small traditional stores in which they were accustomed to shop. Instead, they walked on paving or platforms near the entrance and faced counters behind which salesmen standing on tatami mats hawked their wares. One can still find a few such stores today. Although Mitsukoshi's tatami mat floors were also unsuitable for shoes, customers had to enter the store to shop. And enter they did, many thousands each day.

At the entrance a check room attendant took charge of customers' shoes and handed them slippers to use on the fragile floors of the store. As the number of customers grew, so did the strain on this system. One day five hundred shoes were misplaced, and the historian of Tokyo, Edward Seidensticker, speculates that this disaster may have slowed acceptance of Western methods of distribution until after the 1923 earthquake, when wooden floors were finally introduced.

Mitsukoshi's evolution tells us something we should know by now about technology: it is not merely a means to an end, a neutral tool, but reflects culture, ideology, politics. In this case, two very different nationally specific techniques of flooring competed as an apparently unrelated change occurred in shopping habits. Neither wooden nor tatami mat floors can be considered technically superior, but each has implications for the understanding of "inside" and "outside" in every area of social life, including, of course, shopping. It eventually became clear at Mitsukoshi that Western methods of distribution required Western floors.

The conflict between these flooring techniques has long since been resolved in favor of Western methods in most public spaces in Japan except traditional restaurants, inns, and temples, where one still removes one's shoes before entering. Nevertheless, the tatami mat conserves a powerful symbolic charge for the Japanese, and many homes have both *washitsu* (Japanese-style rooms) and *yoshitsu* (Western-style rooms). This duality has come to seem emblematic of Japan's cultural eclecticism. Globalization there has largely meant conserving aspects of traditional Japanese technique, arts and crafts, and customs alongside an ever-growing mass of Western equivalents. At first it seemed that a Western branch had been grafted onto the Japanese tree. Today one may well ask if it is not a Japanese branch surviving precariously on a tree imported from the West.

This story illustrates the idea of nationally specific branching development. Branching is a general feature of social and cultural development. Ideas and customs circulate easily, even among primitive societies, but they are realized in quite different ways as they travel. Although technical development is constrained to some extent by a causal logic, design is underdetermined, and a variety of possibilities is explored at the inception of any given line of development. Each design corresponds to the interests or vision of a different group of actors. In some cases, the differences are quite considerable, and several distinct designs coexist for an extended period. In modern times, however, the market, political regulations, or corporate dominance dictate a decision for one or another design. Once the decision is consolidated, the winning branch is black boxed and placed beyond controversy and question.

It is precisely this last step that did not take place in the relations between national branches of design until quite recently. Poor communications and transport meant that national branches could coexist for centuries, even millennia, without much awareness of each other and without any possibility of decisive victory for one or another design. Globalization intensifies interaction between national branches, leading to conflicts and decisions such as the one exemplified by Mitsukoshi's floors.

However, conflict and decision are not the only consequence of a globalized world. Here is a second story that illustrates a different pattern I call "layered" development (Malm 1971).

Shortly after the opening of Japan to the world, the Satsuma domain hired a British bandmaster, William Fenton, to train the first Japanese military band. Fenton noticed the lack of a Japanese national anthem and set about creating one. He identified a poem, which is still sung as the lyrics of the Japanese national anthem, and set it to music. This unofficial anthem had its debut in 1870, but it was nearly unsingable and quickly fell into disuse.

The need for an anthem was especially pressing in the navy. Japanese officers were embarrassed by their inability to sing their own anthem at flag ceremonies at sea. The navy therefore invited court musicians to train the navy band in traditional Japanese music in hopes that among the performers a composer would be found. But the process was too slow, and the navy finally asked the court musicians themselves to supply it with a suitable composition. The results were again disappointing. The court musicians came up with a piece in an ancient mode arranged for performance by a traditional ensemble, hardly the sort of thing one would expect to find ready and waiting on a navy ship!

Around this time, Fenton was replaced by a German bandmaster, Franz Eckert. Herr Eckert rose to the occasion. He arranged the anthem supplied by the court for a Western band, making suitable modifications for playability. In 1880, Japan finally had its current national anthem.

This story is quite different from the Mitsukoshi one. Like flooring, music had developed in Japan and the West along different branches; however, the Japanese national anthem is neither Japanese nor Western but draws on both traditions. The relations between traditions in this case are quite complex. The very idea of a national anthem is Western. An anthem is a self-affirmation that implies the existence of others before whom the national self is affirmed. But there were no others for Japan during its 250 years of isolation in a world unto itself. With the opening of the country, self-affirmation became an issue, and an anthem was needed. But how could the anthem affirm Japan unless it reflected Japanese musical style? Hence the composition had to be Japanese. This was easier said than done since the anthem was to be performed by Western instruments at Western-inspired ceremonies. Thus an original Japanese compositional layer had to be overlaid in the final stage with a further Western layer.

Here we do not have rooms of different styles side by side but a true synthesis. The merging of traditions takes place in a layering process that is characteristic also of many types of social, cultural, and technological development. Often several branches can be combined by layering the demands of different actors over a single basic design. In the process what appeared to be conflicting conceptions turn out to be reconcilable after all. The anthem sounds vaguely Japanese played by a brass band. Similarly, modern Japanese politics, literature, painting, architecture, and philosophy emerged in the Meiji era out of a synthesis of native and Western techniques and visions.

Layering should not be conceived on the model of political compromise, although it does build alliances between groups with initially different or even hostile positions. Political compromise involves trade-offs in which each party gives up something to get something. In technological development, as in musical composition, indeed, wherever creative activities have a technical basis of some sort, alliances do not always require trade-offs. Ideally, clever innovations get around obstacles to combining functions, and the layered product is better at everything it does, not compromised in its efficiency by trying to do too much. This is what the French philosopher of technology, Gilbert Simondon, calls "concretization" (Simondon 1958). Concretization gives rise to global technology, combining many national achievements in a single fund of world invention.

The Globalization of Development

Branching and layering are two fundamental developmental patterns. Their relations change as globalization proceeds. Elsewhere I have described two styles of design corresponding to different stages in this process. "Mediation-centered design" characterizes the earlier stage, in which each nation develops its technology relatively independently of the others.[1] The overwhelming weight of particular national traditions ensures that ideas, even ideas of foreign origin, will be incorporated into devices differently in different contexts. These differences are owing in large part to nationally specific ethical and aesthetic mediations that shape design. Thus each design expresses the national background against which it develops.

Globalization imposes a very different pattern I call "system-centered design." The globalizing economy develops around an international capital goods market on which each nation finds the elements it requires to construct the technologies it needs. This market moves building blocks such as gears, axles, electric wires, computer chips, and so on. These can be assembled in many different patterns.[2]

The capital goods market is such a tremendous resource that once interchange between nations intensifies, it quickly becomes indispensable. But when design is based on the assembly of prefabricated parts, it can no longer so easily accommodate different national cultures. Instead of expressing a cultural context, products tend more and more to be designed to fit harmoniously into the existing system of parts and devices. Accommodation to national culture still occurs, but it shares the field with a systematizing imperative that knows no national boundaries. Meanwhile, national culture expresses itself indirectly in the contribution it makes to innovation on the capital goods markets.

The shift toward system-centered design has implications for the role of valuative mediations in the structure of modern, globalized technology. Traditional technologies generally fit well together. Japanese tatami mat floors, traditional architecture, eating and sleeping habits, and shoes all are of a piece. As such, they express a definite choice of way of life, a framework rooted in Japanese culture. However, on purely technical terms, the links between these artifacts are relatively loose. It is true that Japanese houses need entryways in which to leave shoes, that futons must be spread on tatami mats, and so on, but adapting each of these artifacts to the others is not very constraining. The wide margin for choice makes it easy for cultural values to install themselves in technical design.

The globalization of technology changes all this. When design is system based, it must work with very tightly coupled technical components. Electric wires and sockets cannot be designed independently of the appliances that will use the electricity. Wheels, gears, pulleys, and so on come in sizes and types fixed by decisions made in their place of origin. A device using them must accommodate the results of those decisions.

System-centered design thus imposes many constraints at an early stage in the design process, constraints that originate in the core countries of the world system. These constraints are imposed on peripheral nations participating in the globalizing process without regard for their national cultures. Furthermore, the very availability of certain types of capital goods reflects the technological evolution and priorities of the core countries, not those of later recipients. Thus the effect of globalization is to push aside cultural constraints, if not to eliminate them altogether. The products that result appear to be culturally "neutral" at first sight, although in fact they still embody cultural assumptions that become evident with wide use in peripheral contexts.

The computer is an obvious example. For us Westerners, the keyboard appears technically neutral. But had computers been invented and developed first in Japan, or any other country with an ideographic language, it is unlikely that keyboards would have been selected as an input device for a very long time. For the same reason that the fax machine prospered first in Japan, where it facilitated communication in Chinese characters, so computers would probably have been designed early with graphical or voice inputs of some sort. The arrival of Western computers in Japan was an alienating encounter, a challenge to the national language. Considerable cleverness had to be invested in domesticating the keyboard to Japanese usages (Gottleib 2000).

These observations indicate the weakness of national culture in a globalizing technological system; however, there is another side to the story. Countries far from the core, such as Japan used to be, may not contribute as much as core countries, but they do contribute something. And these contributions will be marked by their national cultural background. In the case of Japan, the magnitude of these contributions has grown to the point where they are significant for the original core countries. Global technology contains a Japanese layer and so exhibits a true globalizing pattern, not simply core/periphery relations of dependence.

It is difficult to give examples of this feedback from national culture. The technical realization of a cultural impulse looks just like any other technical artifact. Still, a hermeneutic approach ought to be able to find cultural traces in the technical domain.

Perhaps miniaturization could be cited as a specific contribution reflecting Japanese culture. At least this is the argument of O-Young Lee, whose book *Smaller Is Better: Japan's Mastery of the Miniature*, argues that the triumph of Japanese microelectronics is rooted in age-old cultural impulses (Lee 1984). The practice of miniaturization characteristic of bonsai, haiku poetry, and other aspects of Japanese culture appears in technical artifacts too. Lee cites the early case of the folding fan. Flat fans invented in China arrived in Japan in the Middle Ages. The folding fan, which seems to have been invented in Japan soon afterward, was exported back to China, inaugurating a familiar pattern. The basic technology of the transistor radio and the videotape recorder both came from the United States, but the miniaturization of the devices, which was essential to their commercial success, took place in Japan, from which they are exported back to the United States.

Of course once capital goods markets are flooded with miniaturized components, every country in the world can make small products without cultural afterthoughts. But if Lee is right, the origin of this trend would lie in a specific national culture. Aspects of that culture are communicated worldwide through the technical specifications of its products.

Nishida's Theory of the Global World

In the first part of this chapter I have illustrated a thesis about the globalization of technology with stories from Japan. In the remainder I will try to draw out the implications of this thesis for the major contribution of Japanese philosophy to the understanding of globalization, Nishida's prewar theory of the global world.

Nishida's argument was formulated in the context of the growing self-assertion of Japan in the early twentieth century. For many Japanese, this was primarily a matter of national expansion, but for intellectuals like Nishida the stakes were world cultural leadership. These two aspects of Japan's rise were connected but not identical. On the one hand, Japan had become powerful enough to conquer its neighbors. On the other hand, this very fact showed that Japan, an Asian nation, could participate fully in cultural modernity, assimilating Western achievements and turning them to its own purpose. Nishida argued on this basis that Asia could finally take its place in the modern world as the cultural equal, or even superior, of the West (Nishida 1991, 20).

The link between Nishida's position and Japanese imperialism is thus complex and controversial. I have already contributed to that debate in several articles and will return briefly to this topic in the conclusion of this chapter (Feenberg 1995, chap. 8). However, my main interest here lies elsewhere, in the parallel I find between the structure of technological globalization as I have explained it earlier and Nishida's conception of a "global world" (*sekaiteki sekai*) (Nishida 1965c, 291–292, 294). I will show that the contrast between branching and layering underlies this conception, although Nishida did not develop the technological implications of his own approach.

Nishida argued that until modern times, the world had what he called a "horizontal" structure, that is, it consisted of nations lying side by side on a globe that separated rather than united them. The concept of "world" was necessarily "abstract" during the long period that preceded modern times. By this Nishida meant that "world" was just a concept, not an active force in the lives of nations. This condition was unusually prolonged in the case of Japan, which remained disconnected from growing world commerce and communication until the 1860s.

International commerce transformed this horizontal world by bringing all the nations into intense contact with each other. The result was the emergence of what Nishida called a "vertical" world, a world in which nations struggle for preeminence. Every nation now participates actively in the life of its neighbors through trade, and the movement of people and ideas. There is no harmonious fusion here but rather a hardening of identities that leads ultimately to war. In this context, nationalism emerges as a survival response to the threat of foreign domination.

Nishida had several suggestive terminologies for this shift. He presented a multiplicity of conceptual frameworks, each one inadequate by itself to describe social reality but able to do so all together in a mutually correcting system of categories. The complexity of Nishida's argument is supposed, therefore, to correspond to the actual difficulty of thinking global sociality.

Nishida developed the contrast of horizontal and vertical worlds in terms of the relation of the "many" to the "one" in space and time. The many nations dispersed in space enter into interaction in the modern world. Interaction in history implies more than the mechanical contact of externally related things. Each nation must "express" itself in the world in the sense of enacting the meanings carried in its culture. This can lead to conflict as nations attempt to impose their own perspective. But interaction also requires commonality. Two completely alien entities cannot interact. At each stage in modern history a common framework is supplied by a dominant nation that defines itself as a unifying "world" for all the others. The unification involves the imposition of a general form on the struggles of the particular nations. Nishida gave the example of Great Britain's imposition of the world market on the nineteenth century (Nishida 1991, 24). The many conflicting nations are thus bound together at a deeper level in one world.

The passage from the many to the one is also reflected in the relations of space and time. The dispersal of the nations in space, their "manyness," is complemented by the simultaneity of their coexistence in a unifying temporal dimension. The struggles between the nations have an outcome that is this unity. Thus in modern times, geography is subordinated to history. The unifying nation represents time for this world and as such loses itself in the process of unification it imposes. Britain is absorbed into the world market it creates and becomes the scene on which the world economy operates. The particularity of the nation, Britain, is transcended by the universal order it institutes.

The mechanical and the organic form yet another terminological couple Nishida explores. The mechanical world is made of externally related things dispersed in space. Mechanically related things can properly be called "individual." Their multiplicity forms an "individual many" (*kobutsuteki ta*) (Nishida 1991, 29–31). The organic world consists of wholes oriented toward a *telos* in time. The whole is thus a subject of action, a "holistic one" (*zentaiteki ichi*) (Nishida 1991, 37–38). Society is not adequately described as mechanical because it forms a whole, and yet it is not organic because its members are fully independent individuals, not a herd. The undecidability of the mechanical and the organic indicates the originality of the social world, which cannot be represented by either concept because it embraces aspects of both.

Nishida introduced the concept of "place" (*basho*) in a final attempt to conceptualize this "self-contradictory" globalized world. Place in Nishida's technical sense of the term is the "third" element or medium "in" which interacting agents meet. But a separate entity would itself require a place to interact with the actors. The *basho* is thus not something external to the interaction but a structure of the interaction itself. This structure arises as each actor "negates itself" to become the "world" for the other, that is, the place of the interaction (Nishida 1991, 30).

It is not easy to interpret this obscure formulation. It seems to mean that in acting, the self becomes an object for the other. But the self is not just any object but the environment to which the other must react in asserting itself as subject. As the other reacts, it defines itself anew, and so its identity depends on the action of the self. But the determination of the other by the self is only half the cycle; the action of the other has an equivalent impact on the self. Interaction is the endless switching of these roles, a circulation of self-transforming realizations (*jikaku*) achieved through contact with an other self (Tremblay 2000, 99–101) .

Nishida had two ways of talking about the role of place in the modern world. Sometimes he wrote as though the globalizing nation serves as the "place" of interaction for all the other nations, the scene of interaction. This place can be imposed by domination or freely consented as cultural supremacy, the difference Nishida assumed between England in the past and Japan in the future (Nishida 1991, 99, 77; Nishida 1965c, 373, 349). At other times, he claimed that the modern age is about the emergence of global place in the form of a world culture of national encounter (Nishida 1970, 78–79, 134–135; Ohashi 1997). Nishida did not see any contradiction between these two discourses because he assumed that Japanese culture is a kind of "emptiness" capable of welcoming all cultures. But as we will see, this ambiguity turns out to be quite important.

On the basis of this analysis, Nishida asserted the importance of all modern cultures. Western dominance is only a passing phase, about to give way to an age of Asian self-assertion. The destiny of the human race is to fruitfully combine Western and Eastern culture in a "contradictory self-identity." This concept refers to a synthesis of (national) individuality and (global) totality in which the emerging world culture is supposed to consist.

There is a sense in which this global world constitutes a single being that changes through an inner dynamic. Thus the world "determines itself." But the identities of the particular nations are not lost in this unity. The resulting world culture will not replace national cultures. Something more subtle is involved. Nishida wrote, "A true world culture will be formed only by various cultures preserving their own respective viewpoints, but simultaneously developing themselves through global mediation" (Nishida 1970, 254). World culture is a pure form, a "place" or field of interaction, and not a particularistic alternative to existing national cultures. They persist and are a continuing source of change and progress. The process of self-determination is thus free in the sense of being internally creative; it is not determined by extrinsic forces or atemporal laws. There is nothing "outside" the world that could influence or control it. Even the laws of natural science must be located inside the world as particular historically conditioned acts of thought (Nishida 1991, 36).

Here is a passage in which Nishida describes the global world as he envisaged it:

Every nation/people is established on a historical foundation and possesses a world-historical mission, thereby having a historical life of its own. For nations/peoples to form a global world through self-realization and self-transcendence, each must first of all form a particular world *in accordance with its own regional tradition*. These particular worlds, each based on a historical foundation, unite to form a global world. Each nation/people lives its own unique historical life and at the same time joins in a united global world through carrying out a world historical mission. (Nishida 1965a, 428; Arisaka 1996, 101–102)

However, this cosmopolitan argument culminates strangely in the claim that Japan is the center of the unifying tendency of global culture. Just as Britain unified the world through the world market in the spirit of utilitarian individualism, leading to endless competition and strife, so Japan will unify the world around its uniquely accommodating spiritual culture, leading to an age of peace. Japan will be the "place" on which the world will move beyond the limits of the West to become truly global. Japan can lead the world spiritually because its unique culture corresponds to the actual structure of the global world: "It is in discovering the very principles of the self-formation of the contradictory self-identical world at the heart of our historical development that we should offer our contribution to the world. This comes down to practicing the Imperial Way and is the true meaning of 'eight corners under one roof'" (*hakkoo ichiu*) (Nishida 1991, 70).

The vagueness of this conclusion is disturbing. Nishida explicitly condemned imperialism and argued that Japan cannot be the place of world unity if it acts as a "subject" in conflict with other nations. Instead, it must "negate itself" and become the "world" for all other nations (Nishida 1991, 70, 77). Yet he also recognized the fatal inevitability of world conflict and accepted Japan's military role within that context, as in this statement from his speech to the emperor: "When diverse peoples enter into such a world historical (*sekaishiteki*) relation, there may be conflicts among them such as we see today, but this is only natural. The most world historical (*sekaishiteki*) nation must then serve as a center to stabilize this turbulent period" (Nishida 1965b, 270–271). And, as we saw earlier, he employed ultranationalist slogans, apparently in the hope of being able to instill new meaning into them. The least that one can say is that his efforts were naïve and lent backhanded support to an imperialistic system that conflicted with his own philosophy.

But just as one can question the depth of the connection between Nazism and Heidegger's thought, Nishida's nationalism is similarly ambiguous. There is no clear logical connection between his claims about Japan and his conception of global unity. At least the British gave the world the world market around which to unify. What did Japan have to offer? What mediation did it provide that qualifies it as the center of the new age?

So far as I can tell, Nishida was not bothered by this question, although he should have been. He claimed that Japan is the *archetype* of global unity through its ability to assimilate both Eastern and Western culture, but while this is indeed admirable, it is not clear how it qualifies Japan as the *place* of global unity. For that to be true, Japan would have to do something more positive on the world stage than simply to exist as a model. Nishida does announce the world historical significance of the liberation of Asia from Western imperialism. Yet this is certainly not the equivalent of the world market as a unifying force but rather one of its divisive consequences. In the end this question remains unanswered.[3]

Despite its flaws and limitations, Nishida's theory of globalization remains truly interesting. Nishida claimed that the world has moved from a horizontal to a vertical structure, from indifferent coexistence in space to mutual involvement in time in a conflictual but creative process of global unification. The emerging unity does not efface national differences but incorporates them into an evolving world culture that is best defined as a "place" of encounter and dialog. A common underlying framework makes possible the communication of nations amid their conflicts.

This claim precisely parallels the analysis of the passage from branching to layered development presented in the first part of this chapter. The various branches of technology in a spatially dispersed world finally meet in the global world of modern times. There they assert themselves and come into conflict, but there they also inform each other with ideas and inventions drawn from diverse national traditions. The outcome, global technology, forms a sort of place in Nishida's sense, a scene on which the encounter between nations proceeds without eliminating the originality and difference of the constitutive national cultures. The layering process, in which each culture expresses itself while at the same time contributing to a single fund of invention, is thus precisely congruent with Nishida's conception of world culture.

Japanese Philosophy of Technology

Might Nishida have concretized his approach through a reflection on technology? He came close to making that connection. He understood that historical action is inextricably intertwined with technical creation. He explained that "Culture includes technique" (Nishida 1991, 61). Technique is an expression of a people's spirit as it interacts with the environment and through that interaction forms itself (Nishida 1991, 57; Nishida 1965c, 328). "We create things through technique and in creating them we create ourselves" (Nishida 1991, 33; Nishida 1965c, 297). Although Nishida did not do so, one can build on these observations and carry them a step further by relating this social conception of technique to his notion of global cultural interaction in the twentieth century (Murata 2003, 232–235).

This was what one of Nishida's most brilliant students attempted in a major contribution to philosophy of technology that is all but unknown in the West. Kiyoshi Miki was an unorthodox Marxist who became a Japanese nationalist during World War II.[4] He was influenced by Nishida's teaching and may well have influenced Nishida's views on history and technology cited earlier. In 1939 he published his major work, *The Logic of the Imagination*, in which he explains society as the product of the form-giving power of the imagination. Technology plays a central role in this process as an expression of imagination in the world.

From this standpoint, technology cannot be explained on purely scientific grounds. It lies at the intersection of science and culture, causality and teleology. In his *Philosophy of Technology* (originally published in 1942) Miki wrote,

Truly new inventions not only employ new means, but also create new ends. An inventor should not be thought of as merely inventing new means. . . . Thus machines do not follow the principle of causality alone. Of course, they do follow it to the extent that science is a foundation of technology. But at the same time, they are also teleological. Machines, in their construction and function, embody teleology. . . . [T]echnology can thus be conceived as a unity of causality and teleology. (Miki 1967, vol. 7, 309–310)

Technology is "subjective-objective." It is subjective in the sense that it requires human reason, creativity, and "logos," while also objective in the sense that it manifests itself in a concrete form that confronts us as independent, tangible reality. History is nothing other than this "movement" through technological creation. "As formative action, our actions are historical. Historical actions are technological. Indeed, history is created technologically; historicity cannot be conceived apart from technology" (Miki 1967, vol. 7, 211).

But if this is true, then technology must be imbued with the forms of the culture that created it. And in fact Miki argued that the technology Japan has received from the West is an expression of Western culture and must be reshaped to conform to the Japanese "spirit." A new culture must be created that combines the best of both East and West. Like Nishida, Miki believed the solution to this problem to be of world-historical importance. The West had reached a dead end that Japan could surpass. "The new culture," he wrote, "must on the one hand be rooted in the Eastern tradition with its superb spiritual elements, and on the other hand, it must respect modern technology based on the modern science developed in the West" (Miki 1967, vol. 7, 319).

What was to be the content of this new culture? Miki observed that the Greek term "technē" applied equally to what we call art and technology. This is appropriate since both involve subjective and objective elements. The split between them in modern times is artificial. "The principle of the new culture must rest on [a combination] of the technological and aesthetic worldviews.... [The] 'organicization' of technology is also an 'aestheticization'" (Miki 1967, vol. 7, 329).[5]

One can imagine a development of this approach in terms of Nishida's concept of self-creation through the mediation of a nationally specific technological culture. But Miki did not follow his line of argument to its logical conclusion. Instead he shifted focus from the synthesis of art and technology to a reliance on social science. Rather like Dewey, he argued that the crisis of Western technology would be overcome through a better understanding of society. The outcome of this attempt to create an authentically Japanese philosophy of technology is thus rather disappointing. Miki ended up with a peculiar mixture of nationalism and pragmatism that in its underlying structure might just as well have been invented in Chicago. And like Dewey's similar views, Miki's argument appears far less persuasive after several generations of unsuccessful efforts to create a social science capable of fulfilling the mission they assigned it.

Nishida and Miki witnessed the modernization process as it unfolded in Japan. They were surrounded by rapid social, cultural, and technological change, which they welcomed and which they believed could become the medium for the expression of an authentic Japanese spirit. They rejected the ultranationalist insistence on keeping the Japanese branch pure in the age of global interaction and insisted that Japan should enter the world scene and move forward. In this they were the theorists of their moment in history, a moment in which Japan appeared to be successfully combining Eastern and Western styles in every domain of life. Nishida and Miki lived these events intensely. Perhaps they lost their shoes at Mitsukoshi. Surely they sang the national anthem and were swept along with their country by the syncretic modernization of Japan's government, cities, schools, and cultural production. I conjecture that this background underlay their conception of the global world and their confidence in the future.

Conclusion: Technology and Values

Nishida and Miki attempted to reconcile the specificity of their Japanese ways with the new material framework of life imported from the West. They gave this problem general philosophical significance through an original conception of the imbrication of technology and culture. Their reason for doing so was the belief that Western hegemony and with it Western culture had reached a historical limit. The achievements of the West would now be absorbed into a new historical era organized around Asian culture. In the shadow of Western decline a new sun was rising that would reinterpret the nature of rationality itself. Of course this expectation was disappointed. Japan did not found a new era of Asian supremacy. But the discovery of the socio-cultural contingency of Western technology was to have significant echoes many years later around different issues.

The Japanese case is a subset of the more general problem of the relation of values to technical rationality that we now face around issues such as environmentalism and surveillance. Threats from technology must be balanced against the democratic potential of user agency that has become visible in the development of the Internet. Both threats and potential have brought home to us the same puzzling phenomenon that confronted these Japanese thinkers. Like them, we are confronted with the paradoxical particularity of supposedly universal technical achievements.

Their puzzlement, like ours, is due to Enlightenment assumptions about the nature of rationality. The notion of a rational civilization was first proposed in a polemic context in opposition to traditional religious and feudal beliefs. The opposition of reason and "superstition," that is, the authority of the past, is so fundamental to modern self-understanding that there is no way to break with it that does not carry a risk of regression. However, applied to technology, this rigid dichotomy is misleading. Technology is no pure realization of rationality, but, as we have seen, it condenses technical and social aspects. These dual aspects of technology require a style of analysis and critique quite different from the Enlightenment approach. Reason, at least in its technical realizations, is not universal but just as particular as any other expression of culture.

The Japanese encountered a nationally specific bias when they imported our technology. We are more likely to experience technical bias around issues of access and usage or outright dangers and injustices. This parallel enables us to place the issues raised in the earlier part of this chapter in a wider context.

Branching development corresponds to a world in which the trace of values appears clearly in design features of technical artifacts. Indeed, in traditional technical systems, there is no sharp distinction between technical insights and what we would call "ethical and aesthetic values." There is a "right way" to do things, and it includes all these factors in a single set of practices.

This is the world that Enlightenment rationalism criticized and overthrew. In so doing it freed economic and technical development from restraints laid down in traditional culture. The modern era opens with the struggle for freedom in this peculiar sense of the term. For centuries, progress in the West meant eliminating valuative mediations from rationalized institutions to the greatest extent possible, and this was confused with the emergence of pure rationality from an inheritance of irrational restrictions and limitations.

But as we have seen, values enter into technical choices in other subtler ways that were invisible to Westerners but immediately obvious once Western technology was transferred to Japan. Western technology was shaped by systematizations that were the bearers of a way of life installed in its very design, and this way of life was quite different from the Japanese way. The supposed purity of technical rationality required no elaborate demystification in Japan because it was obviously false. Nishida's and Miki's culturalist interpretation of technology made perfect sense in this context and anticipated the conclusions of contemporary technology studies. The synthesis of Eastern values and Western technology they imagined has its parallel today in the layering of technology with environmental, democratic, and other objectives excluded from the original design process.

Today as the West confronts the limitations of its own technology, it is as though the whole world has begun to resemble modernizing Japan. We are threatened by our technology in ways we can no longer ignore and confronted with our own responsibility and unsuspected powers in a startling reversal of common assumptions. The threat is systematic and resists the familiar modes of critique we have deployed against superstition since the Enlightenment. New ways of understanding and criticizing technology are necessary to enable us to separate the rational core of our technological achievements from undesirable aspects that might be eliminated under a different political dispensation. The growth of a technical public sphere opens new possibilities for democratic interventions into technical development. Philosophy of technology takes on its full significance in this unprecedented situation.

III Modernity and Rationality

The first chapter of this part addresses the relation of modernity theory and technology studies. Both treat the impact of technology on society, the one in terms of the general process of rationalization, the other in terms of the development of specific devices, yet there is no communication between them. This chapter attempts to explain this peculiar disconnect and concludes with an attempt at synthesis around common hermeneutic approaches.

The second chapter explores the sense in which modern societies can be said to be rational. Social rationality describes systems and institutions that bear some resemblance to commonplace notions of rationality such as mathematical equivalence. Markets are socially rational in this sense. This chapter develops a critical strategy for addressing the resistance of social rationality to rational critique. This strategy first appears in Marx's analysis of capitalist economics. The theory of surplus value relies on a conceptual framework similar to the notion of underdetermination in contemporary science and technology studies. But somewhere along the way the critical thrust was diluted. Critical theory of technology attempts to recover that thrust. Here its approach is generalized to cover the three main forms of social rationality.

The third and concluding chapter addresses the central theme of this book: the relation between everyday experience and technological rationality. In traditional societies no great divide separates the realms of knowledge and experience, but in modern times specialization and differentiation have prevailed, and no common culture joins the *disjecta membra*. Heidegger and Marcuse identified this condition with modern technology. Marcuse proposed a radical transformation of technology through joining technical and aesthetic insight. His formulations are ambiguous and have been widely misunderstood. A clarification of his argument leads to a broader reflection on the reform of technical disciplines and their relation to the lessons of experience.

7 Modernity Theory and Technology Studies: Reflections on Bridging the Gap

Posing the Problem

Theories of modernity and technology studies have both made great strides in recent years but remain disconnected despite the obvious overlap in their concerns. How can one expect to understand modernity without an adequate account of the technological developments that make it possible, and how can one study specific technologies without a theory of the larger society in which they develop? These questions have not even been posed, much less answered persuasively, by most leading contributors to the fields. The basic issue I would like to address is the why and wherefore of this peculiar mutual ignorance.[1]

In the first half of this chapter, I will review the positions of some of the major figures in each field. After posing the problem briefly in this section, I sketch the background to the current impasse in the original contributions of Marx and Kuhn. I shall then consider the obstacles each field places in the way of encountering the other. In the second half of the chapter I propose one possible resolution of the dilemma, bridging the gap between the two fields through a synthesis of some of their main contributions. Both modernity theory and technology studies employ hermeneutic approaches, which I elaborate further in a loosely Heideggerian account of innovation. In the concluding sections I summarize my own instrumentalization theory and show how it can be applied to the computerization of society.

Modernity theory relies on the key notion of rationalization to explain the uniqueness of modern societies. "Rationalization" refers to the generalization of technical rationality as a cultural form, specifically, the introduction of calculation and control into social processes with a consequent increase in efficiency. In exposing the traditional social world to technical manipulation, rationalization also reduces its normative and qualitative richness. Modernity theories often claim that this reduction impoverishes experience and unleashes violent forces. But, the theorists argue, impoverished and threatening though it may be, technical rationality gives power over nature, supports large-scale organization, and eliminates many spatial constraints on social interaction. This ambivalent view of modernity is characteristic of a normative style of cultural critique that is anathema to contemporary technology studies. Albert Borgmann's theory of the "device paradigm" is a brilliant example of this approach (Borgmann 1984).

Rationalization depends on a broad pattern of modern development described as the "differentiation" of society. This notion has obvious applications to the separation of property and political power, offices and persons, religion and the state, and so on. But a rationality differentiated from society appears to lie beyond the reach of social study. If technology is a product of such a rationality, it would escape socio-cultural determination.

Technology studies rejects this whole approach. It eschews general theories and relies instead on case studies to show the social complexity of technology, the multiple actors involved with its creation, and the consequent richness of the values embedded in design. Its principles of symmetry cast doubt on the very idea of pure rationality. From this point of view, modernity theory is wrong to claim that all of society stands under values somehow specific to a science and technology differentiated from other spheres. If technology and society are not substantial "things" belonging to separate spheres, it makes no sense to claim that technology dominates society and transforms its values.

But technology studies loses part of the truth when it emphasizes only the social complexity and embeddedness of technology and minimizes the distinctive emphasis on top-down control that accompanies technical rationalization. This trend depends on the differentiation of institutions such as corporations that wield technical rationality without much regard for workers, traditions, and customs. Limited though that differentiation may be, it nevertheless allows any concrete value or thing to be grasped as a manipulable variable, and this includes human beings themselves. Where traditional craft work expressed vocational investment of the whole personality, modern work organization abstracts deskilled occupations from personal character and growth the better to expose the worker to external controls. Similarly, traditional architecture condensed historical and aesthetic expression with stability and durability, whereas today strictly "utilitarian" construction is the rule. True, other values rush in to fill the vacuum left by the differentiation of the technical sphere—for example, profit—but differentiation is a real characteristic of modernity with immense social consequences.

Is it possible to find some truth in both these positions, or are they mutually exclusive, as they certainly appear to be at first sight? I believe a synthesis is possible but only if the concept of technical rationality is revised to free it from implicit positivistic assumptions. It is this positivism that leads modernity theory into the error of assuming that differentiation imposes a purely rational form on social processes when in fact, as technology studies demonstrates, technology is social through and through.

But we must also find a way to preserve modernity theory's insight into the distinctiveness of modernity and its problems. We need to explain how rationality operates even as it is intertwined with society. This technology, that market, will always be socially specific and inexplicable on the terms of a philosophically purified concept of reason.[2] But they will also be unexplainable without reference to their rational form. In the next section I will sketch the background to the two very different ways of understanding rationality in modernity theory and technology studies.

Science of Society and History of Science

The writings of Marx are surely the single most influential source of theories of modernity. His thought is usually identified with a universalistic faith in progress. At its core there is an intuition he shared with his century, the notion that a great divide forever separates premodern from modern societies. All later contrasts of *Gesellschaft* versus *Gemeinschaft*, organic versus mechanical solidarity, traditional versus post-traditional society, and so on owe something to Marx's canonical formulation of this idea in texts such as the *Communist Manifesto* and *Capital*.[3] After World War II, modernization theory emerged as the chief competitor to Marxism, but it shared Marx's progressive universalism.

The sense of radical discontinuity in Marx's texts involves more than a theory of society. His notion of what Max Weber will later call "rationalization" not only covers the changes in economic and technical systems Weber identifies but also includes a new form of individuality freed from ideology and religion. Individuality in this sense is plain to see in the nineteenth-century novels contemporary with Marx's work, and he assumes its generalization to the lower classes because, under the conditions of modern capitalism, workers have no fixed abode and are not subject to the paternalistic authority of nobles and clerics. As the tectonic plates of culture are thrown into movement by the market, the lower classes are freed from naïve faith in their "betters" and achieve an ironic appreciation of the gaps between ideals and realities. Under these conditions, they gain mental independence and become, in Engels's phrase, "free outlaw[s]" (Engels 1970, 23). Marx's social theory is thus founded not just on cognitive hypotheses but also on the existential irony of this modern individual. His method is fundamentally hermeneutic and demystificatory as well as analytic. This duality explains the contrast between the method of Marx's ideology critique and that of his positive economic theory. It shows up in various guises in modernity theory and is especially clear in Habermas, who, as we will see later in this chapter, employs both hermeneutic and analytic methods to study modern society.

If there is any one figure who has played a comparable role for contemporary technology studies, it is Thomas Kuhn. It is true that the case for Kuhn as a founding father is less clear. Many studies of science and technology avoided the positivistic errors Kuhn criticized before the appearance of *The Structure of Scientific Revolutions* (Kuhn 1962). However, Kuhn's overwhelming success lent philosophical legitimacy to these studies and encouraged others to follow their lead. Nonpositivist historiographic methods triumphed in science studies and subsequently influenced the new wave of technology studies that grew out of science studies in the 1980s. Unlike Marx, Kuhn is perhaps less an origin than a symbol of a radically new approach.[4]

Of course neither Marx nor Kuhn is followed slavishly by contemporary scholars, but we should not be surprised to find that many of their background assumptions are still at work in the most up-to-date contributions to modernity theory and technology studies. I would like to begin by considering several such assumptions that may help to explain the gap between these two fields.

Like all modern historians and social theorists, Kuhn writes somewhere in the long shadow cast by Marx, as can be deduced from the reference to "revolution" in the title of his major book, but his take on history is quite different from Marx's. Kuhn's vision of the past, like Marx's, is shaped by the idea of radical discontinuities in history. But where Marx took for granted the existence of a rationality gradient more or less identical with the level of scientific achievement and capable of transcending particular cultures and ordering them in a developmental sequence, Kuhn deconstructed the very idea of a universal standard of rationality. The demystificatory impulse is still present, but it is directed at the belief in progress characteristic of modernity. Now the ironic glance turns back on itself, undermining the cognitive self-assurance implied in the stance of the *naïve* ironist.

Kuhn's approach had momentous consequences for the wider reception of science studies in the academic world. He showed that there is no one continuous scientific tradition but a succession of different traditions, each with its own methods and standards of truth, its own "paradigms." The illusion of continuity arises from glossing over the complexities and ambiguities of scientific change and reconstructing it as a progress leading straight to us. If we go back to the decisive moments of scientific revolution and examine what actually occurred from the standpoint of the participants, their competing positions, their arguments and experimental results, we will see that the case is by no means so clear.

This practice-oriented approach is neatly captured in Latour's suggestion that science resembles a Janus looking back on its past in an entirely different spirit from that in which it looks forward to the future (Latour 1987, 12). Science, Latour suggests, is a sum of results that "hold" under certain conditions, such as repeated experimental tests. The backward glance shows nature confirming the results of science, while the forward glance presents a very different picture in which the results that hold are called "nature." Looking backward, one can say that the conditions of truth were met because the hypotheses of science were true. Looking forward, one must say rather that meeting the conditions defines what scientists will use for truth. The backward glance records an evolutionary progress of knowledge about the way things are independent of science; the forward glance tells of the sheer contingency of the process in which science decides on the way things are.

I doubt if Kuhn would have appreciated this Nietzschean twist on his original contribution, from which unfortunately he retreated in subsequent writings. Kuhn himself never challenged the notion of modernity or the material progress associated with it. But the point is to offer an interpretation not so much of Kuhn as of his significance on the maps of theory. A critique of Marx is implied in his notion of scientific revolution insofar as the latter did believe that his own work was scientific and, more deeply, that progress in rationality characterizes the institutions and forms of modernity by contrast to earlier social formations. Thus just because Kuhn undermines the pretensions of science to access transhistorical truths, his work also undercuts Marxism and the modernity theory that inherited many Marxist assumptions. From that standpoint, it is clear that Kuhn is in some sense the nemesis of Marx and the harbinger of what has come to be called "postmodernism." And to the extent that much technology studies reflects Kuhn's methodological innovations, it too bears a certain elective affinity for postmodernism or at least for a "nonmodern" critique of Marx's heritage.

The implicit conflict came to the surface in various formulations of postmodernism, but it seemed still a mere epistemological disagreement. Philosophers, sociologists, and scientists engaged in heated debates over the nature of truth, but these debates had only a few echoes in modernization theory, such as Habermas's critique of Foucault. Things have changed now that the conflict has emerged inside the ill-matched couple we are considering here: modernity theory and technology studies. Since no fully coherent account of modernity is possible without an approach to technology and vice versa, the philosophical disagreement now appears as a tension between fields. It is no longer just a matter of one's position on the great question of realism versus relativism but concerns basic categories and methods in social theory.

Consider the implications of technology studies for the notion of progress. If Kuhnian relativism has the power to dissolve the self-certainty of science and technology, then what sense does it make to talk about "rationalization"? In most modernity theories, rationalization appears as a spontaneous consequence of the pursuit of efficiency once customary and ideological obstructions are removed. Technology studies, on the contrary, shows that efficiency is not a uniquely constraining motive of design and development but that many social forces play a role. The thesis of "underdetermination" holds that there is no one rational solution to technical problems, and this opens the technical sphere to these various influences. Technical development is not an arrow seeking its target but a tree branching out in many directions. But if the criteria of progress themselves are in flux, societies cannot be located along a single continuum from the less to the more advanced. Like Kuhn's theory of scientific revolutions, but on the scale of society as a whole, constructivist technology studies complicates the notion of progress at the risk of dissolving it altogether.

In Latour's account, a contingent scientific-technical rationality can gain a grip on society at large only through the social practices in which it is actively "exported" out of the laboratory and into the farms, streets, and factories (Latour, 1987: 249ff). The theorists export their relativistic method as they trace the movements of its object. They dissolve all the stable patterns of progress into contingent outcomes of "scaling up" or controversy. Institutional or cultural phenomena no longer have stable identities but must be grasped through the process of their construction. This approach ends up eliminating the very categories of modernity theory, such as universal and particular, reason and tradition, culture and class, which are transformed from explanations into explananda. One can neither rise above the level of case histories nor talk meaningfully about the essence and future of modernity under these conditions.

Modernity theory suffers disaster on its own ground once it encounters the new approach. If no determined path of technical evolution guides social development toward higher stages, if social change can take different paths leading to different types of modern society, then the old certainties of the theory collapse. One can no longer be sure if such essential dimensions of modernity as rationalization and democratization are actually universal, progressive tendencies of modern societies or just local consequences of the peculiar path of recent Western development. Unless it squarely faces the difficulties, the theory of modernity must become so abstract that this objection no longer troubles it, with a consequent loss of usefulness, or cease to be a theory at all and transform itself into a descriptive study of specific cases. Here are two examples that show the depth of the problems.

System or Practice

Modernity as Differentiation

On the whole, modernity theory either continues to ignore technology or acknowledges it in an outmoded deterministic framework. Most revealing is the extreme but instructive case of Jürgen Habermas. Habermas is one of the major social theorists of our time. Yet he has elaborated the most architectonically sophisticated theory of modernity without any reference at all to technology. This blissful indifference to what should surely be a focal concern of any adequate theory of modernity requires explanation, especially since Habermas is strongly influenced by Marx, for whom technology is of central importance.

Habermas's approach is also influenced by Weberian rationalization theory. According to Weber, modernity consists essentially in the differentiation of the various "cultural spheres." The state, the market, religion, law, art, science, technology each become distinct social domains with their own logic and institutional identity. Under these conditions, science and technology take on their familiar post-traditional form as independent disciplines. Scientific-technical rationality is purified of religious and customary elements. Similarly, markets and administrations are liberated from the admixture of religious prejudices and family ties that bound them in the past. They emerge as what Habermas calls "systems" governed by an internal logic of equivalent exchange. Such systems organize an ever-increasing share of daily life in modern societies (Habermas 1984, 1987). Where formerly individuals discussed how to act together for their own mutual benefit or followed customary rituals and roles, we moderns coordinate our actions with minimal communication through the quasi-automatic functioning of markets and administrations.

According to Habermas, the spread of such differentiated systems is the foundation of a complex modern society. But differentiation also releases everyday communicative interaction from the overwhelming burden of coordinating all social action. The communicative sphere, which Habermas calls the "lifeworld," now emerges as a domain in its own right as well. The lifeworld includes the family, the public sphere, education, and all the various contexts in which individuals are shaped as relatively autonomous members of society. It too, according to Habermas, is subject to a specific rationalization consisting in the emergence of democratic institutions and personal freedoms. However contestable this account of modernity, something significant is captured in it. Modern societies really are different, and the difference seems closely related to the impersonal functioning of institutions such as markets and administrations and the increase in personal and political freedom that results from the new freedom of communication.

At first Habermas argued that system rationalization threatened technocratic intrusions into the lifeworld of communicative interaction, and this reference seemed to link his theory to the theme of technology familiar from the first generation of the Frankfurt School (Habermas 1970; Feenberg 1995, chap. 4). However, his mature formulation of the theory ignores technology and focuses exclusively on the spread of markets and administration. The arbitrariness of this exclusion appears clearly in the following summary of Habermas's theory:

> Because we are as fundamentally language-using as tool-using animals, the representation of reason as essentially instrumental and strategic is fatally one-sided. On the other hand, it is indeed the case that those types of rationality have achieved a certain dominance in our culture. The subsystems in which they are centrally institutionalized, the economy and government administration, have increasingly come to pervade other areas of life and make them over in their own image and likeness. The resultant "monetarization" and "bureaucratization" of life is what Habermas refers to as the "colonization of the life world." (McCarthy 1991, 52)

What became of the "tool-using" animal of the first sentence of this passage? Are its only tools money and power? How is it possible to elide technological tools in a society such as ours? The failure of Habermasian critical theorists even to pose much less respond to these questions indicates a fatal weakness in their approach. But there is worse to come.

Habermas's reformulation of Weber's differentiation theory neutralizes rational systems by reducing them to their functional role. This has conservative political implications. In many of Habermas's formulations, for example, when he considers workers' control, radical demands are treated as dedifferentiating, hence as irrational (Habermas 1986, 45, 91, 187). He thus offers no concrete suggestions, at least in *The Theory of Communicative Action*, for reforming markets and administrations and instead suggests limiting the range of their social influence.

In the case of science and technology, this puzzling retreat from a social account is carried to the point of caricature. Habermas claims that science and technology are based quite simply on a nonsocial "objectivating attitude" toward the natural world (Habermas 1984, 1987, I, 238). This would seem to leave no room at all for the social dimension of science and technology, which has been shown over and over to shape the formulation of concepts and designs. Clearly, if scientists and technologists stand in a purely objective relation to nature, there can be no *philosophical* interest in studying the social background of their insights. On Habermas's view, it is difficult to see how a properly differentiated rationality could incorporate social values and attitudes except as sources of error or extrinsic goals governing "use." This also implies a problematic methodological dualism in which a quasi-phenomenological account of the lifeworld coexists with objectivistic systems-theoretic explanations of markets and administrations. No doubt there are objects best analyzed by these different methods, but which method is suited to analyzing the interactions between them? Habermas has little to say on this score beyond his account of the boundary shifts and goal setting that are the only possible points of intersection in his framework.

The effect of this approach is to liberate modernity theory from all the details of sociological and historical study of actual instances of rationality. No matter what story sociologists and historians have to tell about a particular market, administration, or, *afortiori*, technology, this is incidental to the philosophically abstracted forms of differentiated rationality. The real issue is not whether this or that contingent happening might have led to different practical results because all that matters to social theory is the range of rational systems, the extent of their intrusions on the proper terrain of communicative action (Feenberg 1999, chap. 7).

Could it be that the most important differentiation for Habermas is the one that separates his theory from certain sociological and historical disciplines, the material of which he feels he must ignore to pursue his own path as a philosopher? But when the results are compared with earlier theories of modernity, it becomes clear that he pays a tremendous price to win a space for philosophy. Marx had a concrete critique of the revolutionary institutions of his epoch, the market and the factory system, and later modernization theory foresaw a host of social and political consequences of economic development. But Habermas's complaints about the boundaries of welfare state administration seem quite remote from the main sources of social development today, the response to environmental crisis, the revolution in global markets, planetary inequalities, the growth of the Internet and other technologies that are transforming the world. In his work the theory of modernity is no longer concerned with these material issues but operates at a higher level, a level where, unfortunately, very little is going on.

Of course some social theorists have made contributions to the theory of modernity that do touch on technology interestingly.[5] Ulrich Beck has proposed a theory of "reflexive modernity" in which the role of technology is discussed in terms of transformations in the nature of social rationality (Beck 1992). Beck starts out from the same Weberian concept of differentiation as Habermas, but he considers it to be only a stage he calls "simple modernity." Simple modernity creates a technology that is both powerful and fragmented. The uncontrolled interactions between the reified fragments have catastrophic consequences.[6] Beck argues that today a "risk society" is emerging, especially noticeable in the environmental domain. "Risk society . . . arises in the continuity of autonomized modernization processes which are blind and deaf to their own effects and threats. Cumulatively and latently, the latter produce threats which call into question and eventually destroy the foundations of industrial society" (Beck 1994, 5–6).

The risk society is inherently reflexive in the sense that its consequences contradict its premises. As it becomes conscious of the threat it poses for its own survival, reflexivity becomes self-reflection, leading to new kinds of political intervention aimed at transforming industrialism. Beck places his hope for an alternative modernity in a radical commingling of the differentiated spheres that overcomes their isolation and hence their tendency to blunder into unforeseen crises. "The rigid theory of simple modernity, which conceives of system codes as exclusive and assigns each code to one and only one subsystem, blocks out the horizon of future possibilities. . . . This reservoir is discovered and opened up only when code combinations, code alloys and code syntheses are imagined, understood, invented and tried out" (Beck 1994, 32).[7] I will return to this suggestion in the concluding chapter.

This revision of modernity theory is daring and suggestive, but it still rests on a notion of differentiation that would surely be contested by most contemporary students of science and technology. Their major goal has been to show that "differentiation"—Latour calls something similar "purification"—is an illusion, that the various forms of modern rationality belong to the continuum of daily practice rather than to a separate sphere (Latour 1993, 81).

Yet the main phenomena identified by the theory of modernity do certainly exist and require explanation. A puzzling impasse is reached in the interdisciplinary relationship around this problem. Practice-oriented accounts of particular cases cannot be generalized to explain the systemic character of modernity, while differentiation theory appears to be invalidated by what we have learned about the social character of rationality from science and technology studies. A large part of the reason for this impasse, I believe, is the continuing power of disciplinary boundaries that, even where they do not become a theoretical foundation as in Habermas, still divide theorists and researchers. Far from weakening, these boundaries have become still more rigid in the wake of the sharp empiricist turn in science and technology studies and the growing skepticism in this field with regard to the theory of modernity in all its forms. I will turn now to two examples from technology studies to illustrate this point.

The Logic of Symmetry

The constructivist "principle of symmetry" is supposed to ensure that the study of controversies is not biased by knowledge of the outcome (Bloor 1991, 7). Typically the bias takes the form of an "asymmetrical" evaluation of the two sides of the controversy, ascribing "reason" to the winners and "prejudice," "emotion," "stubbornness," "venality," or some other irrational motive to the losers. A similar bias is also presupposed by such basic concepts of modernity theory as rationalization and ideology, which correlate with their normatively marked opposites, irrational social arrangements and scientific knowledge. These concepts appear to be canceled by the principle of symmetry.

Social constructivists' main concern is to achieve a balanced view of controversies in which rationality is not awarded as a prize to one side only but recognized wherever it appears and in which nontechnical motives and methods are not dismissed as distortions but taken into account right alongside technical ones as normal aspects of controversies. The losers often have excellent reasons for their beliefs, and the winners sometimes prevail at least in part through dramatic demonstrations or social advantage. The principle of symmetry orients the researcher toward an even-handed evaluation by contrast to the inevitable prejudice in favor of the winners that colors the backward glance of methodologically unsophisticated observers.

But there is a risk in such even-handedness where technology is concerned: if the outcome cannot be invoked to judge the parties to the controversy, and if all their various motives and rhetorical assets are evaluated without prejudice, how are we to criticize mistakes and assign responsibility? Consider, for example, the analysis of the *Challenger* accident by Harry Collins and Trevor Pinch (Collins and Pinch 1998, chap. 2). Recall that several engineers at Morton Thiokol, the company that designed the space shuttles, at first refused to endorse a cold weather lift-off. They feared that the "O-rings" sealing the sections of the launcher would not perform well at low temperatures. In the event they were proven right, but management overruled them, and the launch went ahead with disastrous results. The standard account of this controversy is asymmetrical, opposing reason (the engineers) to politics (the managers).

Collins and Pinch think otherwise. They show that the O-rings were simply one among many known problems in the *Challenger*'s design. Since no *solid* evidence was available justifying the canceling of the fateful flight, it was *reasonable* to go forward and not a heedless flaunting of a prescient warning. Scheduling needs as well as engineering considerations influenced the decision not because of managerial irresponsibility but as a way of resolving a deadlocked engineering controversy. It appears that no one is to blame for the tragic accident that followed, at least in the sense that this is a case where normally cautious people would in the normal course of events have made the same unfortunate decision.

But the evidence could have supported a rather different conclusion had Collins and Pinch evaluated it in a broader context. Their symmetrical account obscures the asymmetrical treatment of different types of evidence within the technical community they study. It is clear from their presentation that the controversy at Morton Thiokol was irresolvable because of the imperative demand for quantitative data and the denigration of observation, even that of an experienced engineer. Can an analysis of the incident abstain from criticizing this bias?

Roger Boisjoly, who was most vociferous in arguing against a cold weather launch, based his warnings on the evidence of his eyes. This did not meet what Collins and Pinch prissily define as "prevailing technical standards" (Collins and Pinch 1998, 55). The fact that Boisjoly was probably right cannot be dismissed as a mere accident. Rather it says something about the limitations of a certain paradigm of knowledge and suggests the existence of an ideological problematic masked by the principle of symmetry. Could it be that Boisjoly's observations were dismissed and quantitative data demanded mainly to keep NASA on schedule? Or put another way, would the need for quantitative data have seemed compelling in the absence of that pressure? By assimilating this case to every other known risk in the design, without regarding Boisjoly's observations as a legitimate reason for extra caution, Collins and Pinch appear to surrender critical reason to so-called prevailing technical standards.[8]

Now, I cannot claim to have made an independent study of the case, and Collins and Pinch may well have stronger reasons for their views than appear in their exposition.[9] However, we know from experience that quantitative measures are all too easily manipulated to support established policy. For example, quantitative studies were long thought to "prove" the irrelevance of classroom size to learning outcomes, contrary to the testimony of professional teachers. This "proof" was very convenient for state legislators, anxious to cut budgets, but resulted in an educational disaster that, like the *Challenger* accident, could not be denied. Similar abuses of cost/benefit analysis are all too familiar. How can critical reason be brought to bear on cases such as these without applying sociological notions such as "ideology," which presuppose asymmetry?

A related problem bedevils science studies around the supposed opposition of local and global analyses. Science studies scholars sometimes claim that a purely local analysis extended to ever-wider reaches suffices in the study of society without the need for "ungrounded" global categories. This is to be sure a puzzling dichotomy. If the local analysis is sufficiently extended, does it not become nonlocal, indeed global? Why not just generalize from local examples to macrocategories, as does modernity theory?

For Bruno Latour, the analysis of contingent contests for power within specific networks suffices and the introduction of terms such as "culture," "society," or "nature" would simply mask the activities that establish these categories in the first place. "If I do not speak of 'culture,' that is because this word is reserved for only one of the units carved out by Westerners to define man. But forces can only be distributed between the 'human' and the 'nonhuman' locally and to reinforce certain networks" (Latour, 1984: 222–223, my translation).[10] Latour continues in this passage to similarly reduce the terms "society" and "nature" to local actions.

This "symmetry of humans and nonhumans" eliminates any fundamental difference between them. The "social" and the "natural" are to be understood now on the same terms. Attributions of social and natural status are contingent outcomes of processes engaged at a more fundamental level. But then the distinctions we make between the status assigned to such things as a student protest in Paris and a die-off of fish in the Mississippi, a politician's representation of American farmers and a scientist's representation of nuclear forces are all products of the network to which we belong, not presuppositions of it.[11]

This stance appears to have conservative political implications since in any conflictual situation the stronger party establishes the definition of the basic terms, "culture," "nature," and "society," and the defeated cannot appeal to an objective essence to validate their claims *quand-même*. Hans Radder argues that actor-network theory contains an implicit bias toward the victors (Law 1989; Radder 1996, 111–112). John Law's well-known network analysis of Portuguese navigation appears to confirm this as it ignores the fate of the conquered peoples incorporated into the colonial network.

Underlying Latour's difficulty with resistance is the strict operationalism that works as an Ockham's razor, stripping away generations of accumulated sociological and political conceptualization. Social and political theory have been shaped by the historical advance of various oppressed groups from powerlessness to full participation in social and political life. If nature and society are exhaustively defined by the procedures through which they emerge as objects, it is unclear how unsuccessful competitors for the defining role can gain any grip on reality at all, even the feeble grip of ethical exigency. For example, the aspiring citizens of an aristocratic society may wish to appeal to "natural" equality against the caste distinctions imposed by the "collective" to which they belong. But if nature is defined by the collective, not simply ideologically or theoretically but in reality, their appeal would be groundless. Or consider demands for justice of the weak and dominated. The concept of justice stands here for an alternative organization of society, haunting the actual society as its better self. But what can ground the appeal to such transcendent principles if the very meaning of society is defined by the forces that effectively organize and dominate it?

I have argued elsewhere that without a global social theory, it is difficult to establish what I call the "symmetry of program and antiprogram," that is, the equal analytic value of the principal actors' intentions, more or less successfully realized in the structure of the network, and those of the weaker parties they dominate (Feenberg 1999, chap. 5). In particular, the symmetry of humans and nonhumans blocks access to the central insight of modernity theory: the extension of technical control from nature to humans themselves. Although the empiricist preference for the local sounds innocent enough, in excluding all explanations based on the traditional categories of social theory, such as class, culture, ideology, and nature, truly rigorous localism blocks even-handed study of social conflict.

The operational reduction of society and nature seems paradoxically to eliminate the contingency of the phenomena. The case resembles artistic production. A musical composition depends on the composer's decisions, which might have been different, yet once it has been completed it is perfectly self-defined. There is no higher authority to which one might appeal against it. Beethoven's Fifth is a necessary product of the contingencies of its creation. Similarly, Latourian networks define themselves as necessary in the course of their self-creation, with no higher authority able to cast doubt on that definition. The contrary hypothesis, that nature is not simply what the collective takes it to be and that society overflows the bounds imposed on it by those with influence and power, would seem to violate Latour's operationalism. Yet without some such hypothesis, one inevitably ends up in the most uncritical conformism. How can he accept such a hypothesis without his theory cracking open at the seams?

Latour's book on political ecology attempts to address such criticisms (Latour 1999). He faces up to the challenge of explaining oppositional agency, that is, resistance to the dominant definition of the network in which the subject is enrolled. Political morality requires that he find a place for such resistance in his theory. But consistency requires that he do so without reintroducing a transcendent nature or morality.

Latour finds a way of having his operationalist cake and eating it too. He argues that the necessary conditions of opposition can be met without positing transcendent principles. The solution is again operational: look not to the transcendent objects but to the contestatory procedures by which the collective is challenged and transformed. These procedures can prevent premature totalizations or closures that ignore the agency of the weak and violate human rights. In sum, Latour substitutes a democratic doctrine of legitimate debate for nature and morality as the ultimate ground of resistance (Latour 1999, 156, 172–173).

But this solution is ambiguous. Latour's claim might be interpreted as an antitechnocratic constitutional principle: "Thou shalt not interrupt the collective conversation with authoritative findings." He might be saying that this is all that philosophy can persuasively claim without prejudging the content of democratic discourse. On the terms of contemporary political philosophy, this would imply a distinction between the right and the good, the one universally valid, the other contentious and rationally undecidable. That interpretation still leaves open the possibility that ordinary actors could legitimately bring forward appeals to a transcendent nature and society. But this does not seem to satisfy Latour. He wants to expel the transcendent objects not only from theory but also from practice. This is a consequence of ontologizing the network, treating it as the actual foundation of the objects it contains. Short of proposing a double discourse, a true one for the theorist and a false one for the masses, Latour is obliged to introduce his theoretical innovations into the collective conversation as an alternative to the outmoded discourse of transcendence.

These theoretical innovations consist in techniques of local analysis that trace the co-emergence of society and nature in the processes of social, scientific, and technological development. Since these processes are historical, what we call "nature" now develops and changes much as does "society." Pasteur's discovery of lactic acid yeast is a great event not only in Pasteur's life but also in the life of the yeast. Latour refers to Whitehead's process philosophy for a metaphysical sanction for the effacement of the difference between nature and society to make room for a third term out of which both emerge (Latour 1994, 212). This is interesting and provocative as philosophy, but can such ideas become generally available to ordinary people as a substitute for the now-disqualified appeal to transcendent grounds for resistance? That promises to be difficult, requiring that common sense itself become Latourian! Presumably the traditional appeal to an existing "nature," for example, natural equality, would give way in a Latourian society to an appeal for a favorable evolution of nature itself. If I have understood him, Latour is confident this will occur, but that seems unlikely (Latour 1999, 32–33). I conclude that his attempt to evade the conformist implications of his position shows more goodwill than practical plausibility.

Now, there is no intrinsic reason why science studies should seek to explode the entire framework of social theory, and not all current approaches lead to such radical conclusions. Yet the tendency to do so is influential in science studies circles. I call attention to it because it takes to the limit a consequence of certain original methodological choices applied to technology and through technology to modern social life. The results are intriguing but ultimately unsatisfactory.

Splitting the Difference

Interpretation and Worldhood

I now want to suggest a partial resolution of the conflict between modernity theory and technology studies. The key point on which I will focus is the role of interpretation in these two fields. Where society is not studied as a law-governed realm of causal interactions, it is usually considered to be a realm of meaning engaging interacting subjects of some kind, for example, subjects of consciousness or language. Interpretive understanding of society is thus an alternative to deterministic accounts, and hermeneutics appears as an explanatory model better suited to society than the nomological approach imitated from physical science.

The place of interpretation in technology studies should be obvious from its critique of determinism. Technologies do not speak unambiguously but must be interpreted, and this fact calls into question their supposedly determining role since they evolve in function of their interpretation. To the social impact of technology there thus corresponds the technical impact of society. This circularity has social ontological implications. Technologies serve needs while also contributing to the emergence of the very needs they serve; human beings make technologies, which in turn shape what it means to be human. This is the "co-construction" of human beings and society.

These circular relationships are familiar from hermeneutics. The famous "hermeneutic circle" describes the paradoxical nature of interpretative understanding: we can understand only what we already, to some degree, understand. A completely unfamiliar object would remain impenetrable.

However, this circularity is not vicious since we can bootstrap our way to a fuller grasp starting out from a minimal "preunderstanding," "like using the pieces of a puzzle for its own understanding" (Palmer 1969, 25).

Pinch and Bijker's analysis of the bicycle highlights the role of "interpretative flexibility" in the evolution of design (Pinch and Bijker 1989). At the origin, the bicycle had two different meanings for two different social groups. That difference in interpretation of a largely overlapping assemblage of parts yielded designs with distinctive social significance and consequence. Pinch and Bijker conclude that "different interpretations by social groups of the content of artifacts lead by means of different chains of problems and solutions to different further developments" (Pinch and Bijker 1989, 42). We saw something similar with the example of the Minitel in chapter 5. Its meaning as either an informational or a communicative device was in play. But this means that there is no stable, pregiven *telos* of technological development because goals are variables, not constants, and technical devices themselves have no self-evident purpose. Clearly, we are a long way here from the old deterministic conception of technology in which changes in design follow from the technical logic of innovation. Meaning is now central.

Interpretation plays an equally important role for modernity theorists such as Habermas and Heidegger. Both thinkers rely on a contrast between scientific-technical rationality and the phenomenological approach to the articulation of human experience. They privilege the everyday "lifeworld" as an original realm within which human identity and the meaning of the real are first and most profoundly encountered. Interpretation rather than natural law prevails in the study of this realm.

For Heidegger, worlds are realms of meaning and corresponding practices rather than collections of objects, as in conventional usage. A world is "disclosed," according to Heidegger, in the sense that the orientation of the subject opens up a coherent perspective on reality. Heideggerian worlds thus resemble our metaphoric concept of a "world of the theatre," or a "Chinese world." Here interpretation is no specialized intellectual activity but the very basis of our existence as human beings (Spinosa et al. 1997, 17).[12]

In his later work Heidegger developed a radical critique of technology for its power to "de-world," that is, to strip objects of their inherent potentialities and reduce them to mere raw materials. This turn in Heidegger's analysis seems to cancel its hermeneutic import since the message of technology is always the same, what Heidegger calls "enframing" (Heidegger 1977). Although this theory of technology is unremittingly negative, some of his followers have attempted to modify it interestingly.

The early Heideggerian concept of the lifeworld has been applied to innovation by Charles Spinosa, Fernando Flores, and Hubert Dreyfus. As we will see, the major focus of their book is on leadership rather than technology, but this turns out to be a correctable error of emphasis. The authors' starting point in any case is the notion of disclosure, which lies at the center of Heidegger's thought. *Disclosing New Worlds* (1997) takes up Heidegger's basic concepts in the context of a theory of history. The problem to which the book is addressed is how disclosive activities change the world we live in, opening us to new or different perspectives and reorganizing our practices around a different sense of what is real and important. The book reviews three types of disclosive practices corresponding to three types of historical actors.

"Articulations" refocus a community on its core values and practices. This is primarily the task of political leaders. As an example, the authors cite John Kennedy's ability to generate enthusiasm for the space race around themes such as the New Frontier. "Cross-appropriations" weave together values and practices from diverse domains of social life in new patterns that alter the structure of our world. This is the work of successful social movements such as MADD (Mothers Against Drunk Driving), which transposed ideas about responsible behavior from the domain of work into the domain of leisure. Finally, and most significantly, "reconfiguration" is the process in which a marginal practice is transformed into a dominant practice. Entrepreneurs are the agents of reconfiguration, which they accomplish through introducing new products that suggest a new style of life. The focus of *Disclosing New Worlds* is not on the products but on the entrepreneurs. Yet the authors write explicitly, "it is the product or service, not the virtuous life-style of the entrepreneur, that makes the world change . . ." (Spinosa et al. 1997, 45).

Although technology studies is not mentioned, the examples illustrate interpretative flexibility nicely. Gillette's successful introduction of the disposable razor is a textbook case. The traditional straight razor belonged to a world in which men cared for and cherished finely made objects. Gillette sensed the possibility of a redefinition of the masculine relation to objects in terms of control and disposability and furthered that change with a new type of razor. In other words, Gillette did not just serve an existing need for sharper razors.[13]

> The entrepreneurial question was, What did his annoyance at the dullness mean? Did it mean that he just wanted a better-crafted straight-edge razor that kept its edge longer? Or did he want a new way of dealing with things? We shall argue that genuine entrepreneurs are sensitive to the historical questions, not the pragmatic ones, and that what is interesting about their innovations is that they change the style of our practices as a whole in some domain. (Spinosa et al. 1997, 42–43)

The concept of style introduced here is a very general feature of worlds relevant to the design of artifacts. We find more precise tools for discussing the reconfigurative work of artifacts in the notions of "actors" and "scripts" in technology studies (Akrich 1992; Latour 1992). In particular the multiplicity of actors identified in many case histories offers a useful corrective to the book's implicit individualism. The bias toward the heroic disclosive power of poets, philosophers, and statesmen, presumed to be in touch with "Being," has been noted in Heidegger and his followers before. Perhaps the overemphasis on entrepreneurs is a modest expression of that bias. In any case, the individualistic emphasis confirms the tendency of modernity theories to abstract from the world of things. But this time there is a difference: for once a theory lends itself to a shift in emphasis to take technology into account because in fact technology is there already at its core. "A *world*, for Heidegger," the authors write, ". . . is a totality of interrelated pieces of *equipment*, each used to carry out a specific task such as hammering in a nail. These tasks are undertaken so as to achieve certain *purposes*, such as building a house. Finally, this activity enables those performing it to have *identities*, such as being a carpenter" (Spinosa et al. 1997, 17).

Instrumentalization Again: De-worlding and Disclosure

We now have two complementary premises drawn from the two theoretical traditions we are attempting to reconcile. On the one hand, the evolution of technologies depends on the interpretative practices of their users. On the other hand, human beings are essentially interpreters shaped by world-disclosing technologies. Human beings and their technologies are involved in a co-construction without origin. Modernity theory addresses the role of differentiated technical disciplines in the "human control of human beings." Technology studies keeps us focused on the essentially social nature of the technical rationality deployed in those disciplines. The hermeneutic perspective builds a bridge between these different approaches.

A synthesis must enable us to understand the central role of technology in modern life as both technically rational in form and rich in socially specific content. This then is the program: to explain the social and cultural impact of technical rationality without losing track of its concrete social embodiment in actual devices and systems. The concept of world disclosure can be helpful here, on the condition that the analysis be pursued not just in terms of the question of style but more specifically in terms of the practical constitution of technical objects and subjects.

The instrumentalization theory attempts to effect such a synthesis. Instrumentalization theory analyzes technology at two levels. The primary instrumentalization is the process of de-worlding inherent in technical action. The materials engaged in technical processes always already belong to a world that must be shattered as they are released for technical employment. The specific de-worlding effect of technical action touches not only the object but also the subject. The technical actor stands in an insulated, external position with respect to his or her objects. We thus distinguish technical manipulation from the reciprocal relations of everyday communication. Philosophical models of instrumental rationality are generally based on this aspect of the technical. It is, for example, central to Habermas's system/lifeworld distinction and Heidegger's critique of enframing.

Most modernity theory identifies de-worlding with the essence of technology without regard for the complexity of the disclosive dimension achieved in the secondary instrumentalization. I conjecture that this is due to two features of the modern technical sphere. On the one hand, technical disciplines themselves incorporate social factors only in a stripped-down abstract form. The most humane of values, for example, compassion for the sick, is expressed technically in objective specifications such as a medical treatment protocol. The fact that the protocol can be followed without compassion suggests that the objective specifications are really self-sufficient, forming a closed universe from which values are excluded. On the other hand, modern technology has been structured around the extension of impersonal domination to human beings and nature in profound indifference to their needs and interests. This type of technical development depends on restricting the range of social considerations that can be brought to bear on design. Thus de-worlding looms especially large in the worlds disclosed in modern societies. These worlds differ from those of premodern societies in that they do not fully cover over the traces of their founding violence through compensatory strategies.

In demonstrating the contingency of technical development, technology studies encourages us to believe in the possibility of other ways of designing and using technology that show more respect for human and natural needs. But an alternative technology is apparently unimaginable from the external perspective of modernity theorists, who are generally innocent of any involvement with the messy and complex process of actual technical development. The theorists fail to recognize that the de-worlding associated with technology is necessarily and simultaneously entry into another world. The problems of our society are not due to technology as such but to the flaws and limitations of the disclosure it supports under the existing form of modernity.

The duality of technical processes is reflected in the split between modernity theory and technology studies, each of which emphasizes one half of the process. De-worlding is a salient feature of modern societies. They are constantly engaged in disassembling natural objects and traditional ways of doing things and substituting new technically rational ways. Focusing exclusively on the negative aspect of this process yields the dystopian critique we associate with thinkers like the later Heidegger. But de-worlding is only the other side of a process of disclosure that must be understood in social terms. Technology studies emphasizes this aspect of the process. The antinomy results from the inherently dialectical character of technical action, misunderstood unilaterally in each case.

Terminal Subjects

I want to conclude these reflections with an example that I hope will illustrate the fruitfulness of a synthesis of modernity theory and technology studies. I have been involved with the evolution of communication by computer since the early 1980s both as an active participant in innovation and as a researcher. I came to this technology with a background in modernity theory, specifically Heidegger and Marcuse, whose student I was, but it quickly became apparent that they offered little guidance in understanding computerization. Their theories emphasized the role of technologies in dominating nature and human beings. Heidegger dismissed the computer as the pure type of modernity's machinery of control. Its de-worlding power reaches language itself, which is reduced to the mere position of a switch (Heidegger 1998, 140).

But what we were witnessing in the early 1980s was something quite different, the contested emergence of the new communicative practices of online community. Subsequently we have seen cultural critics inspired by modernity theory recycle the old approach for this new application, denouncing, for example, the supposed degradation of human communication on the Internet. Albert Borgmann argues that computer networks de-world the person, reducing human beings to a controllable flow of data (Borgmann 1992, 108). The terminal subject is basically an asocial monster despite the appearance of interaction online. But this critique presupposes that computers are actually a communication medium, if an inferior one, precisely the issue thirty years ago. The prior question that must therefore be posed concerns the emergence of the medium itself. One cannot understand the Internet without considering the development of online community as its most characteristic social innovation. I will return to this innovation as it has touched higher education, where proposals for automated online learning have met determined faculty resistance. Meanwhile, actual online education is emerging as a new kind of communicative practice (Feenberg 2002, chap. 5).

The pattern of these debates can be analyzed in terms of instrumentalization theory. The computer simplifies a full-blown person into a "user" in order to incorporate him or her into the network. Users are decontextualized in the sense that they are stripped of body and community in front of the terminal and positioned as detached technical subjects, pure rational actors. At the same time, a highly simplified world is disclosed in which they are faced with menu choices. They are called to exercise initiative in this world.

Positioning and initiative as described here are correlated as primary and second instrumentalizations, interventions that de-world and disclose. "Positioning" is the general term for occupying the specific locus from which technical action is possible: the "driver's seat." So located, the subject finds itself before a "world" of affordances that invites initiatives of one sort or another. The degree of initiative opened up by any particular positioning indicates the freedom allowed the subject in the given technical context.

Approaches based on modernity theory point out the poverty of the virtual world. This appears to be a function of the very radical de-worlding involved in computing. However, we will see that this critique is not entirely correct although there are types of online activity that confirm it, and certain powerful actors do seek enhanced control through computerization. But modernity theorists overlook the struggles and innovations of users engaged in appropriating the medium to create online communities or educational experiments. In ignoring or dismissing these aspects of computerization, they fall back into determinism.

The "posthumanist" approach to the computer inspired by commentators in cultural studies suffers from related problems. This approach often leads to a singular focus on the most "dehumanizing" aspects of computerization, such as anonymous communication, online role-playing, and cybersex (Turkle 1995). Paradoxically, these aspects of the online experience are interpreted in a positive light as the end of the "centered" self of modernity and the emergence of the new, more fluid, and multiple self of the future (Stone 1995). But such posthumanism is ultimately complicit with the humanistic critique of computerization it pretends to transcend in that it accepts a similar definition of the limits of online interaction. Again, what is missing is any sense of the transformations the technology undergoes at the hands of users, many of whom are animated by more traditional visions than one would suspect from this choice of themes (Feenberg and Bakardjieva 2004).

The lifeworld of technology is the medium within which the actors engage with the computer. Processes of interpretation are central there. Technical resources are not simply pregiven but acquire their meaning through these processes. In Latour's language, the "collective" is re-formed around the contested constitution of the computer as this or that type of mediation responsive to this or that actors' program. But under the influence of theorists like Latour, technology studies has become suspicious of the very terms of the actual debates surrounding computerization. Indeed, Latour's principle of symmetry between humans and nonhumans makes it difficult to recognize the contests between control and communication that emerge with innovations such as the Minitel and the Internet. As we have seen in chapter 5, communication functions were introduced by users rather than treated as normal affordances of the medium by the designers of the systems. To make sense of this history, the competing visions of designers and users must be introduced as a significant shaping force.

Consider the struggle over the future of online education (Feenberg 2002, chap. 5). In the late 1990s, corporate strategists, state legislators, top university administrators, and "futurologists" lined up behind a vision of online education based on automation and deskilling. Their goal was to replace (at least for the masses) face-to-face teaching by professional faculty with an industrial product, infinitely reproducible at decreasing unit cost, like CDs, videodiscs, or software. The overhead of education would decline sharply, and the education "business" would finally become profitable. This is "modernization" with a vengeance.

In opposition to this vision, faculty mobilized in defense of the human touch. This humanistic opposition to computerization took two very different forms. Those opposed in principle to any electronic mediation of education had no impact on the quality of computerization but only on its pace. But there were also numerous faculty who favored a model of online education based on human interaction on computer networks. On this side of the debate, a very different conception of modernity prevailed in which to be modern is to multiply opportunities for and modes of communication. The meaning of the computer shifts from an information source to a communication medium, a support for human development and online community. This alternative can be traced down to the level of technical design, for example, the conception of educational software and the role of discussion forums.

These approaches to online education can be analyzed in terms of the model of de-worlding and disclosing introduced earlier. Educational automation decontextualizes both the learner and the educational "product" by breaking them loose from the existing world of the university. The world disclosed on this basis confronts the learner as technical subject with menus, exercises, and questionnaires rather than with other human beings engaged in a shared learning process.

The alternative model of online education involves a much more complex secondary instrumentalization of the computer in the disclosure of a much richer world. The original positioning of the user is similar: the person facing a machine. But the machine is not a window onto an information mall but rather opens up a social world. The terminal subject is involved as a person in a new kind of social activity and is not limited by a set of canned menu options to the role of individual consumer. The corresponding software opens the range of the subject's initiative far more widely than an automated design. This is a more democratic conception of networking that engages it across a wider range of human needs.

The analysis of the dispute over educational networking reveals patterns that appear throughout modern society. In the domain of media, these patterns involve different combinations of primary and secondary instrumentalizations that privilege either a technocratic model of control or a democratic model of communication. Characteristically, a technocratic notion of modernity inspires a positioning of the user that sharply restricts potential initiative, while a democratic conception enlarges initiative in more complex virtual worlds. Parallel analyses of production technology or environmental problems would reveal similar patterns that could be clarified by reference to the actors' perspectives in similar ways.

Conclusion: Toward Synthesis

Let me conclude now by returning briefly to my starting point. I began by contrasting the theoretical revolutions of Marx and Kuhn and promising to bring them together with a method of analysis that would reconcile modernity theory and technology studies. Can a phenomenology of technical worlds achieve a synthesis? Recall that Marx emphasized the discontinuity introduced into history by what has come to be called "rationalization," the emergence of modern societies based on markets, bureaucracies, and technologies. This view seemed to imply a universalism erasing all cultural difference. By contrast, Kuhn, or at least his followers, subverted the notion of progress implied in Marx's vision of an increasingly rational social process and offered us a history subordinate to culture.

I have argued that rationalization describes the generalization of a particular type of de-worlding involved in technical action. That such de-worlding uproots nature and traditional ways is clear. But on this account, rationalization no longer stands opposed to culture as such but appears as a more or less creative expression of it. In practice this means that there may be many paths of rationalization, each relative to a different cultural framework. Rationality is not an alternative to culture that can stand alone as the principle of a social order for better or worse. Rather rationality in its modern technical form mediates cultural expression in ways that can in principle realize a wide range of values. The poverty of the actual techno-culture must be traced not to the essence of technology but to other aspects of our society such as the economic forces that dominate technical development, design, and the media. This insight challenges us to engage in what Terry Winograd and Fernando Flores have called "ontological designing," the self-conscious construction of technological worlds supporting a desirable conception of what it is to be human (Winograd and Flores 1987, 179).

We can fruitfully combine modernity theory and technology studies in an empirically informed, critical approach. The triviality that threatens a strictly descriptive, empirical approach to such humanly significant technical phenomena as experimentation on human subjects, nuclear power, or online education can be avoided without falling into the opposite error of apriori speculation. The alternative—global condemnation, narrow empiricism—is not exhaustive. There are ways of recovering some of the normative richness of the critique of modernity within a more concrete sociological framework. Concepts like "rationality," which technology studies has set out to demystify, can be employed in a new way, and the implicit emancipatory intent of that demystification can be brought to the surface as an explicit goal. Perhaps someday soon the disciples of Marx and Kuhn will be able to lie down together in the fields of the Lord.

8 From Critical Theory of Technology to the Rational Critique of Rationality

Social Rationality

Types of Rationality

Modern societies are said to be rational in a very special sense that distinguishes them from premodern societies. Theories of rationalization and modernization enshrined this distinction at the heart of twentieth-century social thought. Of course modern societies are not rational in a properly scientific sense of the term. But something about the structure of modernity resembles the rationality of the scientific disciplines, and much has been made of this resemblance in the ideologies that justify or criticize it. The question is, "What is the nature of this resemblance?"

One self-congratulatory answer has it that we are more rational than our ancestors because we have achieved scientific knowledge of nature where they had only myths. There is some truth in this but not much. Even in the advanced countries, the most bizarre beliefs persist and flourish. For example, a majority of Americans believe in angels, but this does not prevent them from doing business in an efficient, modern way we think of as rational. In any case, science itself is no longer analyzed on the terms of old positivistic models of pure rational method but is studied today as a social institution. What is more, people were capable of making discoveries and improving technology long before Galileo and Newton. Some sort of nonscientific rationality was involved in premodern technical progress. Finally, it should be kept in mind that rationality is not necessarily good nor even successful. In its social employment, the concept describes a type of practice, not an end in itself nor even a guarantee of effectiveness. Hitler's Germany exhibited a high degree of organizational rationality with consequences both morally evil and instrumentally disastrous.

For all these reasons, scholars no longer accept the old evolutionary notion, crudely formulated by Comte as a succession of religious, metaphysical, and scientific stages in the progress of civilization. Although this notion has become common sense, it vastly overestimates the extent to which science and especially technology are independent of social influences. In reaction against this view, the very concepts of rationality and modernity have become taboo in much contemporary science and technology studies and postmodern critique. This makes for some inconvenient and misleading rhetorical strategies. We may never have been modern or rational in the Comtean sense of the term, but we have certainly been modern and rational in some other sense that remains to be specified adequately. The challenge is to arrive at a new understanding of these concepts that avoids the pitfalls of the evolutionary view.

The current mood affects the evaluation of the Weberian concept of rationalization, which is often dismissed as uncritically rationalistic. Yet this is to misunderstand the most important aspect of Weber's contribution, which in no way depends on an idealized view of reason. Instead, what interested Weber was the increased importance of "calculation and control" in modern organizations such as government administrations and corporations. Weber pointed out that these organizations conform to principles or employ methods involving precision in measurement, accounting, and technical insight. It is true that his concept of "disenchantment" suggests a reason purified of traditional social influences, but new ones emerge with the triumph of modernity. While his framework has evolutionary implications, they are not of the Comtean sort and do not detract from the real significance of his theory (Weber 1958).

In what follows I will develop an approach to rationalization that depends significantly on Weber. However, I am not a Weberian and anticipate that too close an identification with his position will burden my argument with many unwelcome associations. I therefore introduce the term "social rationality" to refer to phenomena Weber treated under the rubric "rationalization." What I retain from Weber is the emphasis on forms of thought and action that bear some resemblance to scientific principles and practices and the role of modern organizations in generalizing those forms in society at large. Many other aspects of Weber's thought, such as his theses on the Protestant ethic or value-neutral research, are not relevant to my argument.

Social rationality in the sense I give the term depends on three main principles:

1. exchange of equivalents,
2. classification and application of rules,
3. optimization of effort and calculation of results.

Each of these principles looks "rational" as we ordinarily understand the term. Calculation is an exchange of equivalents: the two sides of the equals sign are, precisely, equivalent. All scientific work proceeds by classifying objects and treating them uniformly under rules of some sort. And science measures its objects ever more carefully. Business, like technology, is based on optimizing strategies. Social life in our time thus appears to mirror scientific and technical procedures.

Note that the absence of *social* rationality in no way implies the presence of *individual* irrationality, namely, mere prejudice or emotion. That old-fashioned view of premodern attitudes has long since been abandoned for a more nuanced appreciation of other cultures. Wherever there are human beings, one observes more or less rational *individual* behavior and instrumentally effective *collective* behavior. What is distinctive about social rationality is the role of coordination media such as the market (principle 1) and formal organization and technology (principles 2 and 3). Thus while all three principles of rationality are everywhere at work, in modern societies they are implemented by markets, bureaucratic organizations, and technologies on an unprecedented scale. Let's consider this difference in more detail.

• With some exceptions, premoderns generally exchanged gifts or bartered goods, and where markets existed they were fairly marginal (Mauss 1980). Under feudalism, taxation and rents rather than exchange accounted for most of the movement of goods. By contrast, the modern economy is organized around the exchange of money for an equivalent value in goods or labor.

• Traditional societies apply classifications and rules handed down in a cultural tradition. Modern organizations such as corporations and government agencies construct the classifications and apply the rules. This makes for greater flexibility: the system can change overnight rather than evolving slowly as culture changes. It is designed consciously, not inherited from the past (Guillaume 1975, chap. 3).

• Some individuals in every society attempt to make their activities and techniques more efficient, but only in our society is this the primary work of organizations guided by technical or scientific disciplines, and we alone seek constant progress in both efficiency and measurement. What makes this possible is the unusual degree to which modern societies isolate entrepreneurs and innovators from the consequences of their actions for others and the social order (Latour 1993, 41–43).

In sum, a socially rational society is structured by markets, organizations, and technologies around the three principles of rationality. In this it contrasts to regulation by systems of domination and subordination rather than equal exchange, informal cultural classifications and rules rather than formal ones, and traditional rules of thumb rather than carefully calculated optimizing strategies and techniques.

The Social Critique of Reason

As Habermas has pointed out, social rationality has both a technical and a normative dimension. This is particularly clear in the case of the market. In obeying the principle of exchange, markets respect equality in both the mathematical and moral sense: "The institution of the market . . . promises that exchange relations will be and are just owing to equivalence. . . . The principle of reciprocity is now the organizing principle of the sphere of production and reproduction itself" (Habermas 1970, 97). This particular form of "justice" is essential to the survival of capitalism in the world of inequality it creates. The critic who denounces the consequences of the system is silenced, ironically, by the appeal to justice of those who profit from it at the expense of their fellow human beings.

The fact that capitalism is rationally legitimated has important implications for the development of ideology in modern liberal societies. It sets a pattern in which all modern institutions emphasize the rational character of their activities. Science exemplifies the idea of rational community. Rationalized institutions too justify themselves by reference to reasons, although by no means such compelling ones as scientists adduce for their theories. Compelling or not, the mere fact that rational legitimation is considered necessary and useful exaggerates the role of reason in social life.

The appeal to reason is ambivalent. On the one hand, it justifies the system as fair, governed by unchangeable laws, and ruled by impartial experts. On the other hand, it suggests quite different principles of rationality such as reflective critique and uncoerced agreement. These principles can be traced back at least to the ancient Greeks. They underlie the broader notions of rationality invoked by the early Frankfurt School, Habermas, and this chapter as well. These notions of "communicative rationality," as Habermas calls them, are not based on formal similarities to scientific reasoning but rather on the idea of self-knowledge and the pragmatic conditions of rational argumentation and understanding. But communicative rationality has never structured the central institutions of modern societies.

Social criticism of rationality emerged at the end of the eighteenth century when the principles of rationality began to be applied systematically to human beings on a large scale (Foucault 1977). Increasingly, the population appeared as a resource to be efficiently employed by organizations. Markets gradually took precedence over more personal forms of appropriation and exchange. Technology appeared as an independent force as it shed the traditional value systems and institutions that contextualized it in earlier times.

As economic and technical criteria determine more and more aspects of social life, capacities and needs that lack economic and technical significance are devalued. The dominant institutions of earlier times were even more indifferent to the individuals, but innocently so, insignificantly, as outsiders with respect to the inner life of the small communities making up the social world. Now for the first time, a social order begins to be organized down to the last details while the claims of community are weakened by the increased social and geographical mobility of the population. To the extent that the system fails to encompass all aspects of the lives it controls, the individuals become conscious of themselves as distinct from their social identity. The social and the individual stand opposed, or rather the functionalization of the social makes it possible to be an individual in a new sense opposed to all function.

Rationalization calls forth a romantic critique exemplified in the proud claim of Balzac's antihero Vautrin, "I belong to the opposition called life" (quoted in Picon 1956, 114). The image of life versus mechanism reappears constantly in the critique of social rationality, not just in relation to technology but also markets and bureaucracies that appear metaphorically as social machines. This image culminates in the dystopian literature and philosophy of the twentieth century. But romanticism never succeeded in convincing any large number of people to give up the benefits of modernity despite the fact that capitalism, the economic system that generalized social rationality, turned out to be profoundly oppressive and unfair.

Another critique of social rationality stems from Marx. While many contemporary socialists agreed with Proudhon that "property is theft" and hence not an actual exchange of equivalents, Marx dismissed moralizing complaint and analyzed the workings of the market in economic terms. He developed an immanent critique of the contemporary economic theory of exchange. According to this theory, goods were valued by their labor content and traded for the most part in equivalents. The problem Marx confronted was how to explain the inequalities of capitalist society on the basis of this principle without recourse to implausible notions of merit or origin myths such as the social contract.

It is well known how Marx solved this problem with his theory of surplus value. He argued that under the principle of equal exchange, the value of labor power is measured by the cost of its reproduction just like any other commodity. But the productive power of labor is applied during a working day longer than needed to produce goods equivalent to that cost. The difference, surplus value, accrued to the capitalist and generated the observable inequalities without theft or cheating as many socialists supposed. Marx concluded that this exploitative arrangement is a contingent feature of industrial society, which could have been designed differently under a different economic system.

What can still interest us about this theory is not so much the questionable content as the form: the demonstration that rational principles of social organization can yield a biased outcome. Marx showed that capitalists play by the rules of equal exchange, but he then went on to demystify their claim to fairness. He recognized the rationality of the system, thus affirming its coherence at least within certain historical limits while also uncovering its bias, thus separating its technical and normative dimensions.[1]

But even as Western societies gradually absorbed elements of Marx's critique, similar mystifications arose to hide the bias of other rational systems. Technocratic ideology, reinforced by consumerism, depoliticized public issues and presented a smoothly rational face to a society dominated by wealth. These new mystifications are still effective.

Why is it so difficult to develop a critique of the rationality of modern institutions such as markets and technology? Our intuitive sense of bias is shaped by the Enlightenment struggle against a traditional social order based on myths. The critique of that social order identified what I call "substantive bias," bias in social and psychological attitude. Substantive bias designates some members of society as inferior for all sorts of specious reasons such as lack of intelligence, self-discipline, "blood" or breeding, accent and dress, and so on. The Enlightenment questioned these pseudo-reasons as they applied to lower-class males. The false substantive claims of the dominant ideology were demystified and equality asserted on that basis. This approach set a pattern adopted in the critique of discrimination against women, slaves, the colonized, homosexuals, and potentially any other subordinate group.

Marx focused on what was left uncriticized by the contemporary ideologies that claimed to continue the work of Enlightenment, the monumental fact of economic inequality. Since markets are fair, and the element of rational calculation that characterizes them is confounded with our notion of universal, neutral scientific knowledge, economic rationality escapes criticism of its biased consequences. Marx's methodological revolution consisted in circumventing this obstacle through a deeper analysis of the social dimension of this form of rationality. The most fundamental bias of the capitalist system is due not to irrational practices such as those of religion and feudalism but to the particular way in which it implements the rational principle of equal exchange.

I have introduced the concept of "formal bias" to describe such prejudicial social arrangements. Formal bias prevails wherever the structure or context of rationalized systems or institutions favors a particular social group. Marx's economic theory offers a first modern example of the analysis of a formally biased social arrangement.[2]

There are several different types of formal bias. Sometimes it refers to values embodied in the nature or design of a theoretical system or artifact and sometimes to the values realized through contextualizations. I call the first case a "constitutive bias" and the second an "implementation bias." Here are some examples to clarify the distinction.

Constitutive Bias

• Surveillance systems are biased by their very nature. With some exceptions, their effect is to enhance the power of a minority at the expense of a majority, the surveilled.

• A sidewalk the design of which blocks equal access for the handicapped also exhibits a constitutive bias.

• Machines designed to be the right height for children are biased to favor child labor. As we have seen, an argument can be made in a society using such machines that child labor is technically necessary and efficient. But of course we know that the same type of machines could be redesigned for adults.

• Science represents a special case of constitutive bias. As Gerald Doppelt has argued, the constitution of an object of science depends on valuative decisions about epistemic methods (Doppelt 2008). The previous chapter discussed an example of such a decision in the case of the *Challenger* accident.

Implementation Bias

• A test written in the dominant language of a multilingual community may be fair in itself but have discriminatory impacts on speakers of minority languages. In this case there is nothing wrong with the test that could not be corrected by simply translating it.

• Urban plans that concentrate waste dumps near racial minorities are biased by the way in which the dumps relate to a context, not by the fact of their nature or design.

• The digital divide is another case where implementation has discriminatory consequences: it strengthens the rich at the expense of the poor but only because the artifacts are distributed in a specific context of wealth and poverty, not because computers are inherently bad for the poor. In fact they can be a means of social advancement once the poor get hold of them.

Markets and administrations resemble artifacts in that they too can be structured in different ways. Paying women less than men for the same work would be a case of constitutive bias. An auction held at a time and place when legitimate buyers cannot attend would exhibit implementation bias. Basing college admissions on the specific measures that reduce minority admissions is constitutively biased. A city plan that routes freeways through poor neighborhoods rather than rich ones is biased in the implementation.

Marxism and the Politics of Technology

Marx's critical method was not applied to technology in the years following the publication of *Capital*. Marx himself focused primarily on the first principle of social rationality, the exchange of equivalents. But in *Capital* he hints at the class character of technology (Marx, 1906 reprint: I, part IV). The critique of the formal bias of markets is extended here less rigorously to the division of labor and mechanization. For example, Marx writes, "It would be possible to write quite a history of inventions, since 1830, for the sole purpose of supplying capital with weapons against the revolts of the working class" (Marx, 1906 reprint: I, 476; Feenberg 2002, 47–48). He argues here that the form taken by technical progress under capitalism accords with the needs of enterprise rather than society as a whole. It was only in the 1970s that labor process theory recovered this aspect of Marx's thought and brought it up to date (Braverman 1974).

Nineteenth-century socialists were so fascinated by the idea of historical laws that they ignored Marx's critique of technology and focused on his economic theory. While drawing on Marx's notion of modernity, Weber founded the field of organizational sociology on uncriticized capitalist assumptions. He was most interested in the second principle of rationality, classification and the application of rules, as these procedures characterize bureaucratic and business organizations. But Weber lost the Marxian insight into the role of technology and class. Influential successors such as Parsons compounded his error. Nevertheless, Weber's contribution is important as the most successful early attempt to thematize the problem of social rationality as such (Weber 1958). Weber's "iron cage" of bureaucracy is echoed in Lukács's important early work, *History and Class Consciousness*. There Lukács attempted to unify Marx's notion of the "fetishism of commodities" with Weber's rationalization theory in an innovative theory of reification (Lukács 1971).

Lukács provides the link between Marx and the Frankfurt School. Works such as *Dialectic of Enlightenment* (Adorno and Horkheimer 1972) and *One-dimensional Man* (Marcuse 1964) are often dismissed as irrationalist and romantic when in fact they intend a rational critique of a new object. That object, omnipresent technology, is based on calculation and optimization and shapes not just technical devices and social systems but also individual consciousness. Organizations, technologies, and culture are inextricably intertwined, each depending on the others for its design and indeed for its very existence. According to the Frankfurt School, advanced industrial society is "totally administered" as a bureaucratic-technical system.

This extremely negative view of modernity results from a dystopian overemphasis on the limits of agency in socially rational systems. As a result the Frankfurt School often serves as a left-wing version of Heidegger. But in Heidegger the social is completely absorbed into the technical sphere and no longer offers any basis for resistance. In his terminology, nontechnical social forces, were they conceivable under modern conditions, would be merely ontic and subordinated to the ontological fundamentals revealed in the technical functionalization of the world. In contrast, the Frankfurt School proposed a dialectical conception in which the technical and the social are moments in a totality rather than situated in a hierarchy of more and less fundamental.

This is apparent in occasional comments by Adorno and lengthier analyses in Marcuse's work. In one surprising passage that seems to contradict the "critique of instrumental reason" in *Dialectic of Enlightenment*, Adorno writes:

It is not technology which is calamitous, but its entanglement with societal conditions in which it is fettered. I would just remind you that considerations of the interests of profit and dominance have channelled technical development: by now it coincides fatally with the needs of control. Not by accident has the invention of means of destruction become the prototype of the new quality of technology. By contrast, those of its potentials which diverge from dominance, centralism and violence against nature, and which might well allow much of the damage done literally and figuratively by technology to be healed, have withered. (Adorno 2000, 161–162, note 15)

This passage is no more than a promissory note that Adorno never fulfilled, but Marcuse went much further in arguing that technology could be redesigned under different social conditions to serve rather than to dominate humanity and nature (Marcuse 1964, chap. 8). This is the subject of the next chapter.

Although the first generation of the Frankfurt School pursued a version of the Marxian approach under the new conditions of managerial capitalism and state socialism, its formulations are not entirely satisfactory. Ambiguities lend credence to charges of romantic irrationalism. Abandoning Lukács's daring critique and its echoes in the early Frankfurt School, Habermas and his followers avoid all discussion of technology and express open skepticism about workers' control and radical environmental reform. The Habermasians seem to concede that experts can resolve all technical questions properly and appropriately so long as they do not overstep the bounds of their authority and "colonize the lifeworld" (Habermas 1986, 45, 91, 187). With this concession to the autonomy of expertise, they have thrown the baby out with the bath water. And they have done so just when technology has become a major political issue.

Since the 1960s a new politics of technology has gradually refuted the old belief that technical controversies could be resolved through scientific consensus. Instead we have seen the rapid proliferation of lawsuits, demonstrations, and political campaigns over all sorts of technical issues. Students of Marx should not be surprised since many of these conflicts repeat in new arenas struggles similar to those he found in the nineteenth-century factory. Technology has spilled over into every aspect of social life. Medicine, education, games, sports, entertainment, urban design, transportation are all highly technologized, and technology has widespread effects not just on human beings but also on nature. There are controversies and struggles in all these areas, as in the factories Marx studied, over how to organize a "rational" way of life.

Today we no longer expect technical progress to resemble the old image of scientists bending over an experimental apparatus and nodding their heads in agreement. Indeed we no longer believe that even scientists find agreement so simple. Our model of technical advance increasingly resembles ordinary politics. Diverse interests now contend for influence over the design of technologies just as they have always fought for influence over legislation. Each alternative design of medical technologies, transportation systems, the Internet, educational technology, and so on has its advocates whose ideology, way of life or wealth depends on control of technical designs. These controversies appear on the front pages of the newspapers daily as we enter a new era of technical politics.

This is why I have reformulated the Frankfurt School's approach as the "rational critique of rationality" it was intended to be. Recent constructivist technology studies has been useful for this purpose. It is possible to combine insights drawn from the Frankfurt School with recent technology studies because technology studies itself resembles the Marxian critique of social rationality that inspired Adorno, Horkheimer, and Marcuse. Even though most of its practitioners are unaware or unappreciative of Marx's contribution, their own research unwittingly reproduces the very structure of his argument. Technology studies is engaged in a critique of formal bias insofar as it recognizes the political significance of these controversies and struggles.

The generalization of insights from technology studies in the context of a critical theory of social rationality suggests the possibility of radical transformation through political action. However, this hopeful approach requires a theory of social struggle over technological design that neither the Frankfurt School nor contemporary technology studies has developed. The critical theory of technology fills this gap.

Generalized Instrumentalization Theory

The Instrumentalizations

The instrumentalization theory applies not just to technology but also, with suitable modifications, to any socially rational system or institution. They each realize one or more of the three principles of social rationality under specific social, cultural, and political conditions.

Devices are thus situated in two radically different but essentially interlinked contexts: the technical context of rationality and the lifeworld context of meaning. A similar duality is apparent in the spheres of bureaucracy and the economy. The critique of modern society must therefore function at two levels, the level of socially rational operations and the level of the sociocultural conditions that specify definite designs.

As I explained in chapter 4, I call these two levels the "primary" and the "secondary" instrumentalization. The relationship between them is not external: the device does not pre-exist the social determinations of its design. No pure market relations or natural kinds preexist the operations in which markets and classifications are configured. Society and its rational systems are not separate entities. The distinction between them is primarily analytic and methodologically useful. It is not a real distinction between things that exist independently of each other.[3]

I have presented many technological examples in earlier chapters. But the same structures are found in other rationalized institutions. Bureaucracies are configured around systems of classification. These systems reflect the abstraction of what are called "cases" from the concrete flow of the lifeworld. A complex living human situation becomes a case when it is decontextualized and reduced just as a natural object is perceived in terms of affordances in the technical sphere. To pull cases together under a class governed by a rule corresponds roughly to the functionalization of affordances in technical work. And as with technology, bureaucracy loses much of the richness of the lifeworld with the result that tensions arise between it and its clients.

The same illusion of pure rationality appears with bureaucracy as with technology. Classification of such things as crimes, diseases, or educational credentials may permit a bureaucracy to act coherently on what its members take to be an objective basis. Yet numerous rationally underdetermined problems must be solved in the construction of such systems. Often no decisive reason can be adduced to justify one solution over another. In fact classification systems are the result of negotiations, conflicts, and the exclusion of alternatives that might have been brought forward by interested parties too weak to make their voice heard (Bowker and Star 2002, 44).

Similarly, a system such as the market involves operations of equivalence that have a rational character but the framework within which these operations are performed is not itself an exchange of equivalents. Rather, it stems from the social and political conditions governing the market. Those conditions provide the decision rules that resolve underdetermined design choices. An example of such a choice is the boundary of the economy that determines just what can become a commodity and what is excluded from sale and purchase. In chapter 2 I showed how environmental politics is shaped by such considerations.

The socially rational properties of the various systems expose them to mediation by each other. Bureaucratic rationality lends itself to technical mediation, for example, through computerization of clearly and distinctly labeled case files. Similarly, commodification is often supported by technical mediations as in the currently contentious case of the digital watermarking of music and film. The overlapping of modes of rationalization in cases such as these creates an apparently seamless web of instrumentalities.

In the remainder of this section I will sketch the various instrumental-izations that shape objects and institutions in modern societies. These ratio-nalizing processes affect the object, the subject, and cognition.

1 The Object The initial insight that opens up an object to incorpora-tion into a rational system presupposes two conceptual operations. First the object must be decontextualized, split off from its original environ-ment. And second, it must be reduced or simplified to bring to promi-nence just those aspects that can be functionalized in terms of a goal. These operations describe the original imaginative vision of the world in which affordances are identified that expose objects and persons to tech-nization, commodification, and bureaucratic control. For example, a har-vested tree is stripped of its complex connections to other living things and the earth. A person enters the purview of a bureaucracy as a "case," abstracted from the totality of a life process and simplified of extraneous elements. Goods become commodities through an interpretation that strips them bare of human connections and throws them into circulation. "In short, rationalization might be defined as the destruction or ignoring of information in order to facilitate its processing" (Beniger 1986, 15).

Capitalism introduces the most extensive rationalization in history, radically decontextualizing and simplifying a wide range of natural and social elements for incorporation into a system of production and distri-bution. Things treated as raw materials are broken loose from their natural site and stripped down or processed to expose their one useful aspect in the context of production. In the production process they acquire new qualities suiting them to the human context for which they are destined in consumption. People are processed too. They are removed from the tradi-tional domestic work context and relocated in factories. They cannot of course be stripped of their nonproductive aspects like trees or minerals, but they can be obliged by the rules of the workplace to expose only their productive qualities at work.

Commodification is the key operation through which these transfor-mations take place. According to Paul Thompson's account, a good becomes a commodity when:

1. Alienability is enabled (the ability to separate one good from another, or from the person of a human being).

2. There is an increase in excludability (the cost of preventing others from use of the good or service).

3. There is an increase in rivalry (the extent to which alternate uses of goods are incompatible).

4. Goods are standardized (there is an increase in the degree to which one sample of a given commodity is treated as equivalent to any other sample) (Thompson 2006).

Each of the commodification processes can be described under categories of the instrumentalization theory. Alienation and exclusion decontextualize objects, while rivalry and standardization simplify them. Once decontextualized and simplified, objects can be incorporated into a rational system, in this case the market, through appropriate systematizations, for example, by assigning them a distinctive form and a price. So configured, goods and labor circulate on markets, freed from the supposedly "irrational" encrustations on the economy of a traditional society in which religious and family obligations intrude on production.

Passing from the level of the initial functionalization to the actual making of a device or configuration of a market or bureaucracy brings in a host of new constraints and possibilities reflecting the existing technical and social environment. At every stage in the elaboration of a technical device or system, from the original creation of its elements to its final finished form, more and more underdetermined design decisions are made in response to social constraints.

These constraints are of two main types. Before it can be deployed, the decontextualized object must be recontextualized in the framework of a way of life. "Systematization," as I call this process, grants the artifact or other rational system a specific meaning within the system of meanings that constitutes the "world" of the society.[4] On that basis systematizations link the artifact or system to its environment. For example, a way of life that separates work from residence assigns automobiles their place, or research defining "poverty" assigns the social work bureaucracy its role and rules. In addition, the reductions the object has undergone must be compensated by new valuative mediations drawn from the ethical, aesthetic, and other normative registers of the society in which it is to function. These mediations intervene in the design process, determining an object capable of entering a specific social world.

2 The Subject Rationalizing operations are performed by a detached, autonomous subject that is strategically positioned to make use of its objects' causal properties. As Bacon wrote, "Nature to be commanded must be obeyed." The actor's commanding stance has two seemingly contradictory prerequisites. On the one hand, the actor must be able to defer feedback from its action or reduce that feedback in scope; and on the other hand, the actor must obey the independent logic of the system to accomplish an end. Technical examples of the first point are obvious. Hammering in a nail has a big impact on the nail, but the energy that rebounds on the carpenter is of no consequences. Shooting a rabbit may be fatal for the rabbit but has a trivial impact on the hunter, and so on. This is the sense in which the actor can be considered autonomous. Economic examples of the second point are also obvious: as an investor, I do not attempt to change the world but to occupy a market position where the crowd of later investors will find my property and bid up its value.

The capitalist exemplifies the autonomization of the subject in rational systems. The individual capitalist is unlikely to be very different from other people, but insofar as he acts out of a new type of institutional base his practice has a remarkable characteristic: indifference to the social and natural environment within which optimization is pursued. The capitalist as subject thus lacks "humanity" in the traditional sense. This is a detached subject free to a great extent from social control and positioned strategically to make a profit. On this condition it is able to achieve effective technical control of nature, labor, and markets.

But this is not the end of the story. The detached actor finds itself engaged with its objects in a way that determines its identity, and called on to exercise initiative in manipulating them. As noted earlier, the hunter is not much affected physically by killing the rabbit, but his actions designate him as a hunter and as such he takes the initiatives implied in the hunt. The capitalist may be indifferent to each investment and employee, but she is a capitalist with all that that implies. The consumer is detached with respect to each commodity and yet an identity and corresponding activity are shaped by a pattern of consumption. What is deferred at the causal level returns at the level of meaning. This has the practical effect of "configuring" consumers and users and "scripting" their behavior (Woolgar, 1991).

3 Cognition The activities associated with socially rational systems are complemented by cognitive relations that also reflect the two levels of instrumentalization outlined earlier. In suggesting that cognitive relations are so structured, I do not wish to enter an epistemological debate over rationality, but remain at the phenomenological level. At that level, what is at stake is how subjects experience the world, not the nature of truth and reality.

The decontextualized and reduced experience of the initial encounter with affordances involves a perception of or reasoning about causality. The idea of piling stone on stone to build a wall is obviously dependent on causal thinking. But to build up a complex structure such as a house starting out from these simple beginnings actors must integrate a much broader range of experience. That broader range is a world of meanings, a "lifeworld." In every society a house embodies a specific range of meanings assigned it by the culture and this determines design.

The design of modern rational systems is no different from earlier craft techniques in this respect. It must integrate lifeworld meaning and technical insight to be intelligible to members of the society. At the same time, institutionally differentiated technical work and formalized technical disciplines depend on maintaining a certain conceptual distance between functional abstractions and their lifeworld context. This operation is generally absent in premodern societies. It is accomplished by abstracting from valuative mediations to allow complex systematic connections to be elaborated in thought.

The object described in its purely technical aspect can be configured differently in response to different social constraints. This is what gives a sense to the idea of the technical as such. But formal engineering knowledge of these common features is not a device any more than a musical score is a symphony. Similarly, administrative and economic science describe an abstraction, not a reality. The object of formal description does not exist independent of its socially conditioned realization.

Although the technical disciplines abstract from lifeworld contexts, aspects of the secondary instrumentalizations translated into technical specifications appear within them, and reflect an earlier state of society. Other secondary instrumentalizations remain external to these disciplines as discursive expressions of contemporary users and participants aiming at changes in design. These changes may someday become technical standards. Thus considered historically, rational systems are not autonomous but are traversed through and through by the logic of the lifeworld that they shape and that shapes them.

Table 8.1 sums up all the relations involved in the instrumentalization theory as it has been presented earlier.

Function and Meaning

The concept of biological and technical function has the peculiarity of lying at the intersection of causality and teleology. Every such function can be described in both terms: "the purpose of X is Y" is roughly equivalent to some form of "X causes Y." Hence, "the purpose of this switch is to start the engine" could be rephrased to say, "this switch causes the engine to start." Philosophers have argued over the extent of the dependence of different types of function on causal preconditions. Functions established by mere convention, such as the meaning of words, lie at one extreme and such things as hammers and nails at the other. However, much is overlooked in these debates. The emphasis on purpose obscures another aspect of functional objects that I call "meaning." The duality of function and meaning underlies the "double aspects" of the instrumentalization theory.

The distinction between function and meaning is ignored in the recent philosophical literature. Searle, for example, constructs his social ontology around the apparently exhaustive contrast between the natural and the functional qualities of objects (Searle 1995). "Function" refers to any intentional human interaction with a thing. A similar inflation of the concept of function afflicts the interesting contributions of Preston (1998) and

Table 8.1
Instrumentalization Theory (adapted from Feenberg 1999, 208)

	Functionalization	Realization
Objectification	decontextualization reduction	Systematization mediation
Subjectivation	autonomization positioning	identity initiative
Cognitive Relation	causality nature	meaning lifeworld

Kroes and Meijers (2002). They recognize that the range of properties of technical objects is much wider than function in the narrow technical sense (Preston 1998, 246; Kroes and Meijers 2002, 36). Nevertheless, they apply the word "function" in various attenuated senses to all these properties. Of course everything that enters the social process is practically related to human beings, but calling all such relations "functions" is misleading, given their variety, and confusing, given the much stricter notion of function in technical fields.

In these fields, a function is the designated purpose of a bundle of affordances orchestrated in a feature. When technical workers are told what function their work must serve, they look around for materials with affordances that can be combined and bent to this purpose. The secondary instrumentalization intervenes in the realization of the function in features. The affordances must be cast in a form acceptable to eventual users situated in a definite social context. Since technical workers usually share much of that context, many secondary instrumentalizations occur more or less unconsciously. Others are the result of using previously designed materials that embody the effects of earlier social interventions. Still others are dictated by laws and regulations or management decisions. Technical workers are of course aware that they are building a product for a specific user community, and to some degree they design in accordance with an amateur sociology of the user. This sociologizing task may be assigned to others in the organizations for which they work. .

This mutual imbrication of function and meaning distinguishes my approach from Habermas's. In his theory, system and lifeworld represent distinct social spheres. But the differentiation of systems from the lifeworld is nowhere near as complete as he assumes (Feenberg 1999, chap. 7). The routine penetration of systems by lifeworldly meanings shows up in matters of design and configuration that cannot be adequately addressed by systems theory and that receive only the most cursory attention in Habermas's theory of communicative action. Rather than a sharp distinction, a sliding scale of differentiation is indicated, going from the most semantically impoverished to the richest object relations. Meaning is not something extrinsic to the realm of social rationality.

Philosophers overlook meaning in part because the examples they introduce in the discussion of function are usually biological organs or tools. This tends to simplify a complicated picture. Consider a very different type of example such as clothing or table utensils. Clothing has the obvious technical function of protecting the body from the elements, but we know that it also does many other things such as projecting an image of our person, and hiding our nakedness. No doubt these more complex usages can be described as functions too, but to do so misses the point. Projecting an image and hiding nakedness can only be understood in the context of a cultural system, which gives meaning to "image" and "naked." Table utensils have a similar character. Their obvious technical function of moving food from plate to mouth is only an aspect of the ritual usages that surround them. Everything from their design to their appropriate position on the table to the specific task they are assigned is culturally specified. To call each of these aspects a function stretches the term to the point of meaninglessness.

Cultural systems are not reducible to a collection of individual functions because they define a lifeworld within which functions emerge. As such they encompass symbols, feelings, taboos, myths, social structures, and many other things that have only remote connections to what we usually mean by the word "function." Interpreting and describing worlds in this hermeneutic sense is essential to explaining how functional objects are understood and used.

The significance of the distinction between function and meaning is clear in Cowan's sociology of consumption (Cowan 1987) and social histories of technology such as Schivelbusch's (Schivelbusch 1988) study of the industrialization of light, and Armstrong's study of glass in the nineteenth century (Armstrong 2008). In common with other sociologists and social historians, they highlight the hermeneutic complexity of technical change rather than reducing it to a single abstract concept.

The importance of the hermeneutic perspective is evident in the case of the Minitel, discussed in chapter 5. The original function served by the Minitel, the distribution of information, responded not just to a specific need but also to an overall conception of life in a modern society. That conception was at least partially valid. The modern world does pose problems for everyone that can only be solved by quick access to relevant information. To be a successful member of a modern society is to be an information consumer. The communicative subversion of the Minitel responded to a very different conception of modernity in which the atomization of society appeared as a problem at least as important as the need for information. Overcoming social isolation involved not consumption but production, specifically the production of discourse and image shared on the network in chat rooms. These two tendencies combined to break down the traditional separation of the public and private realms, opening the home to the outside world and vice versa. A cultural shift occurred through the appropriation of the Minitel, which continues today on a still larger scale with the Internet.

Much of what is of interest in the story of the Minitel concerns larger social issues and patterns of experience and self-understanding that are not essential to the technical workings of the system. Yet these cultural dimensions did play an essential role in its dissemination and evolution. This is not to say that the Minitel's function was irrelevant to the story. On the contrary, this approach shows how that functional dimension was signified and resignified in its social context.

The links between cultural meaning and function cannot be explained from a functional standpoint. It would be more accurate to say that function is abstracted from meaning, a more complex system of relations in the lifeword.[5] I employ the term "abstraction" here in the Hegelian sense, to refer to taking a part for a complex whole. Function is that aspect of the whole described by "meaning" that is specified technically (or "translated") in features.[6]

But to understand any given functional object or system culturally, it is necessary to undo the work of abstraction and conceive its function as the way in which an aspect of the lifeworld expresses itself in rational form and works itself out. To be sure, an automobile is a means of transportation, but that definition is abstracted from a cultural framework within which space has a particular quality. In that context, automotive transportation signifies the freedom of the individual in a world where residence is separated from work, the distribution of goods, and most other destinations. In sum, the meaning of "transportation" and therefore of "means of transportation" is relative to the lifeworld that determines the spatial distribution of things.

The most puzzling aspect of modern rationality is the existence of purified technical disciplines based on functional abstractions that, despite their purification, continue to interact with the lifeworld from which they have been differentiated. Both sides of this equation must be maintained, difficult though that may be. The demystifying strategy of the main trends in science and technology studies aims to reduce the gap and show that rationality is far less pure than it appears to be. This is an important lesson hammered home in much of the research done in the last thirty years. But just as important is recognition that the abstractions constructed by the technical disciplines are no illusion but are reality-changing interventions into the lifeworld.

Design Codes

Standard ways of understanding and making devices are called "black boxing" in constructivist studies of technology. Many of these standards reflect specific social demands shaping design. In chapter 1 I introduced the concept of the technical code to explain this phenomenon. A similar standardization of design occurs in other socially rational domains. Markets and bureaucracies are more obviously social than technology, but the standards underlying their design tend to be just as invisible. These social standards can be analyzed on the same terms as the technical code. I call the generalized concept referring to the standardization of rational systems the "design code." Design codes are durable, but they can be revised in response to changes in law, economic conditions, public sentiment, and taste.

In this respect, design codes are similar to law in a democratic state. Much democratic politics resembles an institutionalized version of the interactions between initial expert encoding and lifeworld recoding. The modern democratic state is essentially a vast administrative system that is more or less responsive to the lifeworld through the activity of citizens in the public sphere and of their elected representatives in an assembly that mirrors that sphere to some extent. Laws, like design codes, establish stable regularities in social life. Laws depend in the first instance on the identification of classes of phenomena. Such classes are themselves abstracted from lifeworld contexts much as are affordances. Tensions and conflicts emerge where the abstraction leaves behind essential aspects of social life. These tensions may lead to protests and eventually to change, closing the democratic circle.

Design codes are sometimes explicitly formulated in specifications or regulations. But often they are implicit in culture, training, and design, and need to be extracted by sociological analysis. The researcher formulates the code as an ideal-typical norm governing design, but in reality there exist two very different instances of that ideal-type: specifications formulated by experts on the one hand, and expressions of desire and complaints by lay users or victims on the other hand. It is the experts' job to make sure the specifications fulfill lay expectations. This requires a process of translation between a technical discourse and social, cultural, and political discourses. The translation process is ongoing and fraught with difficulty but nevertheless largely effective. This process is made visible in the researcher's ideal-typical formulations of the design code.

The democratic implications of translation are easier to grasp now than in the past. As rational systems intrude on more and more social settings, the resistant lifeworld generates ever-more secondary instrumentalizations. In my earlier work, I verified this dynamic in three domains: online education, human communication on computer networks, and experimental medicine.

In the first case, innovations introduced by lay actors were colonized by computer specialists and commercially oriented administrators. Limitations of the technology and resistance from users have yielded a hybrid system (Feenberg 2002, chap. 5). In the other two cases, a technocratic or scientific ethos presided over the construction of a new environment and in each of these cases lay actors brought to it a self-understanding very different from the designers' expectations. Out of the confrontation of users and technical systems a layered design emerged that served a broader range of human needs than was originally envisaged (see chapter 5; Feenberg et al., 1996; Feenberg 1995, chap. 5).[7] Such changes are democratic and progressive in character. They are essential to maintaining the openness of the rationalized social world.

Conclusion

Modern societies are unique in the exorbitant role they assign social rationality. This has been a significant obstacle to the development of critical consciousness from the earliest versions of free market ideology down to the present technocratic legitimation of advanced societies. It is far more difficult to identify and criticize the formal bias of socially rational artifacts and institutions than inherited mythic and traditional legitimations. A variety of strategies has been tried for this purpose, each growing out of a focus on one or another rationalized institution. The instrumentalization theory is based on critical strategies developed in relation to technology. Here an attempt is made to generalize it to other rationalized spheres.

This brief discussion of the instrumentalization theory can be summarized in the following seven propositions:

1. The theory is a critique of social rationality loosely parallel to Marx's critique of market rationality.

2. The theory is based on analysis of the formal bias of socially rational systems and artifacts.

3. This bias is traced in the seamless combination in design of analytically distinguishable primary and secondary instrumentalizations.

4. Affordances are discovered at the level of the primary instrumentalization with minimal social constraints.

5. These affordances are combined in formally biased systems and devices embodying a wide range of social constraints described in the secondary instrumentalization.

6. Codes determine stable regularities in the design or configuration of socially rational systems and artifacts.

7. Tensions between design and lifeworld contexts give rise to demands that are eventually translated into new codes and designs.

I have sketched here a number of adaptations of the instrumentalization theory to other forms of social rationality, but clearly more work remains to be done. The terms of the instrumentalization theory and the notions of technical code and function must be reconstructed in the different contexts of bureaucracy and the market. A theory of formalization must be developed to explain the relation between technical disciplines and the lifeworld. Other socially rational systems such as games must be studied (Feenberg 1995, chap. 9; Grimes and Feenberg 2009). And a grounded account of the difference between premodern and modern society must be elaborated that avoids both the rationalistic excesses of previous theories of modernization and the polemic rejection of Marxism and the sociological tradition characteristic of much science and technology studies. This is the agenda of a future research program on social rationality.

9 Between Reason and Experience

Introduction

Everyday experience has a teleological character that ancient science raised to the level of an ontological principle. In modern times, the new mechanistic concept of nature shattered the harmony between experience and scientific rationality (Whitehead 2004, 30–31). The world split into two incommensurable spheres: a rational but meaningless nature and a human environment still rich in meaning but without rational foundation. In the centuries since the scientific revolution, no persuasive way has been found to validate experience or to reunite the worlds despite the repeated attempts of philosophers from Hegel to Heidegger. This is not just a theoretical problem. Experience teaches caution and respect for people and things. Experience brings recognition that the Other has its own powers, limits, and goals. Once the lessons of experience no longer shape technical advance, it is guided exclusively by the pursuit of wealth and power. The outcome calls into question the viability of modernity. The genocidal twentieth century is now followed by a new century of environmental crisis.

Technology stands at the crossroads of all these developments. It is both an application of scientific-technical rationality and the background of the world of experience. Communication between the two realms ought to be possible around technical problems if nowhere else. Philosophy of technology thus has a unique vantage point from which to consider the modern dilemma. This vantage point has been occupied fruitfully by Heidegger, whose concept of world is deeply implicated in his notion of technical practice. Yet Heidegger himself failed to draw out the most important implications of this coincidence. Marcuse studied with Heidegger, and although he soon rejected his teacher's doctrine, its subtle influence continued throughout his career. His more socially concrete formulation of a critique of technology opens the way to a new approach that I will sketch in the conclusion of this chapter.[1]

The philosophical issue concerns the relation of norms derived from concrete experience to rationalized technical practice. The expulsion of teleology from scientific-technical rationality stripped it of most normative elements. So long as ethical and aesthetic principles remain external to technique, they appear to intrude impotently on a self-sufficient sphere with its own laws and logic of development. Nothing is more urgent today than rooting these principles in the structure of technical disciplines as restraints on the deployment of their overwhelming destructive power. Can this be accomplished in a progressive framework? Can normativity be restored within the technical realm without regressive re-enchantment of nature or general impoverishment?

These are the questions raised by the thought of Heidegger and Marcuse. It is not easy to recapture the potent force of their criticism in an environment in which many of their ideas have become clichés. Their complex philosophical language makes the task still more difficult. Both Heidegger and Marcuse believe that the question of technology concerns not merely the social problems they criticize but also the very nature of the rational and the real. In order to break through the fog surrounding their ideas, I will begin by constructing a cultural framework of interpretation that I will then apply to explaining their argument. I do not pretend that this framework is adequate as an interpretation of Heidegger and Marcuse but rather will use it to bootstrap from commonsense assumptions to an understanding of their difficult thought.

A Cultural Framework

Culture supplies the meanings that things take on within the social world. It distinguishes our actions from natural events by making it possible for ourselves and others to "read" our meaning and purpose. In another sense, culture bears a significant resemblance to nature. Indeed our most basic cultural assumptions are what we take for nature, the usually unquestioned and unquestionable premises of our thinking, acting, and speaking. For the most part, we operate on the basis of these premises without formulating them consciously.

Cultural assumptions are more stable and widely shared than matters of opinion. But they too can be called into question although always against a background of other assumptions that are not thematized and challenged. There can be no "view from nowhere," from a beyond of all culture. Culture evolves but generally not through direct challenge so much as through gradual changes in practices and taste of which people are scarcely aware. Culture is more or less securely armored against challenge and change, depending on the nature of the social system. A stable and isolated tribal society is more likely to preserve its culture than a rapidly changing modern society in global contact with other modern societies. As a consequence, under modern conditions culture is far easier to question, hence far less "cultural."

In common usage, premodern "craft" is contrasted to modern "technology." Both are ways of making artifacts using tools, but they differ in the scale of their activities and their cognitive basis. Craft employs hand tools in small workshops, whereas modern technology operates at huge scales and has correspondingly huge impacts on nature and society. Traditional crafts serve and express their culture, while our technology is in constant motion, disrupting social institutions and destabilizing cultural life. The difference is in large part a function of the application of scientific and engineering knowledge to which craftspeople did not have access in the past.

While important, these distinctions miss a still more basic difference between the cultural roles of technology and craft. What distinguishes technology most fundamentally is the differentiation of technical activity from other types of social activity. Specifically, technical knowledge is separated out from the prevailing aesthetic and ethical values. The separateness of these categories seems obvious to us. We do not expect technical know-how to involve artistic creativity or building things to involve ethics. But in craft they form a single complex. The craftsperson knows the "right way" to make things, and this involves realizing the "essence" of the artifact in the appropriate materials. Technical knowledge and skill are required, but aesthetic and ethical principles also contribute to the outcome. Without their contribution it is impossible to specify a culturally acceptable artifact. Considerations such as beauty are thus not conceived as subjective values in the head of the craftsperson but as objective facts about the world, like other culturally secured beliefs. Superficial ornamentation, added to artifacts for sales purposes, and similarly motivated packaging are modern inventions that reflect the modern distinction between values and facts.

Max Weber introduced the notion of differentiation that describes the distinctive aspect of modernity. Weber observed the tendency of modern societies to separate functions that were united in earlier times. For example, offices and persons are no longer indissolubly linked in a modern civil service. No longer are social functions inherited, but instead positions are "filled" by qualified personnel. Modernity involves the generalization of such distinctions. Differentiation is more or less complete, depending on the domain.

The differentiation of knowledge of nature from other cultural spheres leads to the development of modern science, based on rational procedures and experiment and validated by an expert community. As science advances, nature is, in Weber's phrase, "disenchanted," stripped of anthropomorphic and spiritual qualities and reduced to a meaningless mechanism. Under this dispensation, science achieves considerable independence of other social institutions.

Something similar happens to technical know-how. It is gradually formalized in technical disciplines that resemble and are enriched by science. This connection gives the impression that technology is just as autonomous as science, but in fact technology is far less differentiated. All technical activity is deeply marked by culture. This is just as true of modern technology as of the crafts of premodern societies. But the mark of culture on technology is much harder to identify, at least for us who belong to the modern world.

In the first place, the cultural context shows up in design. Since modern design emphasizes function, and functions appear self-evident to us, it is easy to overlook its dependence on culture. But cultural limitations become obvious when devices are transferred to alien cultures. In chapter 6 I gave the example of computers with a Roman keyboard exported to Japan, where the language cannot be represented easily by our alphabet. The necessity of adaptation testifies to the cultural relativity of Western computer design.

But there is a more paradoxical way in which modern technology depends on culture: its so-called value freedom. Modern technology falls under the formal norm of efficiency, but efficiency does not determine the particulars of design and use as did the old culturally secured essences. Liberated from such cultural constants, technology can be designed to serve temporary and shifting purposes. This suits it for employment by organizations, another prominent feature of modernity. Like technologies, modern organizations are generally dedicated to rather narrow formal goals such as profitability. These goals are no more able than efficiency to determine any particular outcome of production. For that, the leaders of organizations must rely on their understanding of the contingencies of the market and legal and administrative rules. In the absence of specific cultural direction, these considerations decide what to make and how to make it. Insofar as such decisions lack a stable basis in the culture, technology pursues ends that appear more or less arbitrary. This strange cultural void is itself the culture of technology we hardly question.

To us it appears universal, but it is uniquely compatible with our culture. This is clear, for instance, from Lauriston Sharp's account of the effects of the distribution of steel axes by missionaries in an Australian Aboriginal community (Sharp 1952). The community prized the stone axes made by its adult male members. These axes were not available as pure means in our sense but were bound up with various rituals of ownership and use. Men alone were authorized by the traditions of the tribe to own and lend the axes to women and children for their customary tasks. This system broke down when missionaries distributed steel axes to anyone who helped with the work of the mission.. The social hierarchy, the trade and social relations, even the cosmology of the tribe collapsed, and its members were demoralized. Thus replacing a product of craft by a modern technology implied a profound cultural change and not merely an increase in efficiency.

The criticism of technology to which we are accustomed focuses on the use of technology to achieve particular ends of which we disapprove. We would like to reform the organizations that command the technology and make them serve enlightened purposes. Social movements and state regulation aim to achieve this. But the philosophical critique of technology goes considerably further. Although philosophers do not generally use my sociological terminology, they identify what I have called "differentiation" and the disenchantment of nature that is its consequence as the problem to be addressed.

Insofar as the differentiation of technology belongs essentially to modern culture, this criticism appears strange. Can it be that the philosophers want us to return to the premodern past? Of course not! The reason for their general discontent is not so hard to understand. Modern societies are fraught with meaninglessness, manipulation, and rationalized violence. Dystopia and apocalypse beckon as surveillance and nuclear technologies advance. Climate change melts the poles while nations dither. The long-run survival of modern society is very much in doubt. Could it be that our technology, or at least the specific way in which we are technological, threatens us with early self-destruction? This is the question of the radical critique of technology.

Heidegger's Critique of Technology

Heidegger's critique of technology is ontological, not sociological. Although I have provided the terms of a sociological translation of his argument in the previous section, it is not my intention to substitute sociology for ontology. Rather I hope that the sociological translation will serve as a bridge to understanding his thought. Heidegger's ontology is so contrary to common sense that a bridge is necessary. He is at his most counterintuitive in his dismissal of epistemology. We tend to think that reality is "out there," while our consciousness is an inner domain that gains access to things through the senses. Heidegger rejects this model. He invents his own vocabulary in which terms such as "revealing," "disclosure," "*Dasein*," and "world" substitute for familiar concepts such as "perception" and "consciousness," "culture" and "nature."

As Heidegger explains it, our most basic relation to reality is not perception as we usually understand it. That is a theoretical construction. Abstracting from our actual experience, we tell ourselves about such things as light rays entering the eye and activating the retina, sound waves causing vibrations in our ear drums, and so on. But we originally encounter the world not through causal interaction between nature and the senses but rather through action directed at meaningful objects. We later reflect on these primordial encounters with objects, but Heidegger rejects the notion that we can explain them in a philosophically significant sense from that standpoint. Instead we need to start out from what is first, our actual experience, and treat it as an irreducible ontological basis.[2]

Heidegger argues that the subject of action is not consciousness or the mind but rather what he calls *"Dasein,"* the human being as the site of experience. It is our whole self that engages with reality, not a specialized mental function. Heidegger calls the things that *Dasein* encounters in action "ready-to-hand." This locution refers to the way in which they are given in that specific aspect by which they can be used, the affordances they offer. His examples are tools that we encounter in use through grasping them and setting them to work. In this context we do not focus on their objective properties but rather on the correct way to handle them. Meanings originally emerge out of use and underlie our knowledge of things.

While Heidegger would certainly reject the concept of culture introduced earlier as subjectivistic, it is helpful for understanding his concept of meaning. A hammer is a hammer only insofar as it is culturally signified as such. Outside of any cultural context, it is just an oddly shaped piece of metal and wood. Thus the meaning of the hammer is in fact constitutive of its being a hammer. This is obvious in the case of paper money. A hundred-dollar bill is worth a hundred dollars only because the meaning of money is culturally established. Even a legal definition of the bill would fail if we did not understand the money as money. Heidegger employs a parallel argument in an ontological account of the objects of experience. On this account what is usually called "culture"—shared meanings—is not merely a coincidence of subjective states but founds a world.

In an attempt to avoid any hint of subjectivism, Heidegger substitutes the concept of "worldhood" for the usual concept of experience. *Dasein* is essentially "in" a world of ready-to-hand things. Such worlds are contingent on human concerns without being subjective. Worlds emerge in the human encounter with reality, but that encounter cannot be understood in causal terms because on those terms no world appears but only isolated stimuli and response. "World" must be understood instead as the existential enactment of meaning, not the object of perception. But despite the rejection of a causal account, Heidegger describes the encounter with world in more or less passive terms as a revealing, an opening, not a construction.

Difficult as it is to imagine this encounter, we do have experiences that give a hint of Heidegger's intent. Consider what happens when a museum docent points out the significance of a pattern in a painting. The pattern is "revealed" to us but it would hardly help to understand what we discover to point out that certain rays of light are entering our eyes and being converted into images in the brain. Similarly, when we recognize that a group of symptoms we experience corresponds to a particular illness, acknowledgment of their meaning can be described as a "disclosure" in which both subject and object are involved. Although these are relatively unusual occurrences, Heidegger thinks that something like this goes on in every encounter with meaning. The world is a network of ready-to-hand things disclosed in a system of such meanings. His language struggles to evoke this disclosure, which is taken for granted and indeed must be taken for granted for everyday life to go on.

These are familiar aspects of Heidegger's early thought, but it seems to me that insufficient attention has been paid to the nature of the enactment in which worlds emerge. This relative neglect is, I believe, due to the entwinement of Heidegger's argument from the very beginning with a phenomenological concept of the technical. His work promises a new basis for understanding human life through a radical reevaluation of the structure and relevance of everyday experience in its technical aspect. He challenges philosophy to leave its ivory tower and engage with concrete social reality. For a brief moment existentialism and Western Marxism pursued this path, but they did not succeed in permanently transforming philosophy.

Heidegger's own closest approach to the concrete was disastrous. In 1933 he imagined that he could realize his own philosophy through collaboration with the Nazi regime. The infamous Rectoral Address contains an intriguing ambiguity relevant to my argument. The subject of the address is "*Wissenschaft*" and its place in the university. He hoped to stimulate a reform of the university that would bring its disciplines into a renewed version of the original "Greek" relationship to the world. That relationship he described as one of fearless questioning of reality combined with submission to "fate." Characteristically, he failed to provide any concrete guidelines for accomplishing this in modern Germany. But even while remaining at an ineffectually high level of abstraction, his argument invokes the technical as the domain of decisive ontological encounters.

Heidegger quotes a saying attributed to Prometheus that, he claims, "expresses the essence of knowledge." The text reads "*technē d'anangkes asthenestera makrō*," translated as "But knowledge is far less powerful than necessity" (Heidegger 1993b, 31). Note that Heidegger translates "*technē*" as "knowledge" and thus apparently confounds the know-how of practical making (*technē*) with the *Wissenschaften* (*epistemai*) of university professors. And he insists! In the following paragraphs he rejects the familiar notion that the Greeks idealized disinterested contemplation and writes instead that for them

"theory" does not happen for its own sake; it happens only as a result of the passion to remain close to what is as such and to be beset by it. On the other hand, however, the Greeks struggled to understand and carry out this contemplative questioning as a—indeed as the—highest mode of man's *energeia*, of man's "being at work." It was not their wish to bring practice into line with theory, but the other way around: to understand theory as the supreme realization of genuine practice. (Heidegger 1993c, 31–32)

This obscure formulation must have puzzled his audience. Only his own students would have understood what Heidegger meant by these references to *technē* and *energeia* and this unconventional explanation of Greek science as dependent on practice. In his contemporary lectures he explains that the metaphysical concept of *energeia* signifies the "standing forth" of the worked-up object. *Energeia* is actuality in the sense of the realizing of the essence in the work. The fullest actuality of human beings is the realization of their capacities, their "*dynamis*," in "being at work" in the practice of a *technē*. Heidegger argues that the sciences emerged out of practice at the origins of Greek thought when technical engagement with beings evolved into wonder. Heidegger writes:

it is clear that this perceiving of beings in their unconcealedness is not a mere gaping, that wonder is carried out rather in a procedure against beings, but in such a way that these themselves precisely show themselves. For that is what *technē* means: to grasp beings as emerging out of themselves in the way they show themselves, in their outward look, *eidos*, idea, and, in accord with this, to care for beings themselves and to let them grow, i.e., to order oneself within beings as a whole through productions and institutions. (Heidegger 1994, 155)

In sum, Heidegger appears to be saying that scientific knowledge of the nature of things is not essentially contemplative but grows out of practical craft knowledge.[3] But knowing implies more than making. In knowing, the meaning of what is becomes explicit as idea, essence; it is grasped, Heidegger assures us, in wonder. This respectful attitude lies at the foundation of the sciences and must be recaptured for the university to return to its rightful role in society.

The identification of knowledge and *technē* is familiar from pragmatism, but Heidegger does not reduce truth to consequences. Knowledge is rooted in instrumental activity in the broadest sense but not in that aspect of it that serves mastery of the environment. Technical power worries Heidegger more and more, but at least until the mid-1930s, the form of technical practice has a very broad significance for his philosophy. It is the fact that instrumental activity brings forth something prefigured in an image, an *eidos*, that interests him. In his course on Aristotle's *Metaphysics* he offers such an account of *technē*, explaining the Greek thinker's teleological conception of being as a generalization from craft practice (Heidegger 1994, 76–77).

Heidegger's critique of modern technology contrasts it to this idea of craft. Greek *technē* is an undifferentiated practice. The meanings that underlie it are fixed by the culture so securely that they are not modified or questioned. These meanings are not strictly functional in our modern sense but include what we would call "aesthetic" and "ethical" values as well as technical considerations. The Greeks invented a philosophical terminology in which to refer to the complex meaning in which all these considerations are united, calling it the "essence" of the thing. On the terms of the instrumentalization theory, this concept refers to the undifferentiated primary and secondary instrumentalization conceived as a single entity.

We tend to think of the concept of essence as prescientific, but our artifacts too are often richly signified in much the same way. For example, a house is also a home. Along with the functional good of shelter, it provides welcome and privacy, a locus for the rituals of family life, and a testimony to the taste of the owner. Technological thinking isolates function from this complex, and this attitude is confirmed by the fact that function can be specified in a technical discipline that guides the building of the house. The instrumental dimension appears to be a separate entity, an infrastructure to which superstructural valuative associations are attached. Although it is an abstraction from the totality of the thing, function is substituted for the whole in an ontological synecdoche characteristic of modernity.

By contrast, Heidegger explains the unified structure of essence in terms of the four causes of Aristotle. The final cause is the purpose of the artifact. Its formal cause is the shape it must assume in the course of production.

The material cause is the raw materials. And the efficient cause is the activity of the craftsman. Together they define the work of craft.

This sounds quite commonplace, but we think so, Heidegger claims, only because we misunderstand it on modern terms. He insists that the efficient cause is not a cause in our modern sense at all. The craftsman does not make the object in accordance with his intentions in a relation of cause and effect as modern common sense would have it. Rather the craftsman "gathers" the other three causes and thereby "brings forth" the object of his actions. Craft, Heidegger argues, is a way in which things become what they truly are (Heidegger 1977, 8–10).

What does this rather obscure complication of Aristotle's apparently simple theory really mean? To understand Heidegger's answer to this question, we must shift our focus. As we have seen, for Heidegger, what things are, their essence, consists first and foremost in their meaning. Heidegger thus insists that we view technical making primarily as the realization of a meaning in the world. On this account, every artifact is what it is through conforming with its purpose and form.

This way of thinking about productive activity leads to paradoxical results, at least so they seem to us. The essence is immanent in the practice of making and guides it in the transformation of the materials. The transformation responds to the specific privation the materials suffer in their original condition at the outset of the work. What the material becomes at the hands of the craftsman is not arbitrary but corresponds to a destiny inscribed in its very nature. Heidegger writes, for example, that for the Greeks the potter's clay takes on form under his hands, but, more significantly, it loses its formlessness (Heidegger 1995, 74). It is as though the clay achieves its true end in becoming a pot.

This explains Heidegger's rejection of our modern notion of causality as we would apply it to craft activity. The craftsperson does not make the artifact; he or she lets it become. It enters the world not by a making but by a cultivating, a pruning, a channeling, which enables its inner tendency to realize itself. As Heidegger writes, "The end which finishes, however, is in its essence, boundary, *peras*. To produce something is in itself to forge something into its boundaries. . . . Every work is in its essence 'exclusive' (a fact for which we barbarians for a long time now lack the facility)" (Heidegger 1995, 118). In sum, for the Greeks craft does not create through causal interaction with materials as does modern technology but reveals things that nature unaided cannot bring into the world.

This conception of craft conforms with an old story about Michelangelo. When asked how he made his statue of David, he replied, "I just cut away everything that wasn't David." We feel this to be paradoxical since it presupposes the reality of the statue prior to its actual production, but something like that describes Heidegger's version of the Greek worldview. Like the statue of David, essences in Heidegger's interpretation of the Greeks are realized not so much through a positive act of production as through the exclusion of the inessential, of that which deviates from the nature of the thing awaiting realization. Hence the concept of essence can be thought of as a limit that specifies the thing, the negative of a positive. The Greek view of nature was teleological and attributed essences in this sense not just to artifacts but also to nature. The cosmos was an order created from a primordial chaos by *limitation* (Heidegger 1995, 118). On this Greek model, culture, as a system of meanings, exists through imposing a limit on the infinite possibilities of action and objects.

There is, however, a danger associated with *technē* that Heidegger emphasized increasingly as his work matured. *Technē* "is carried out . . . in a procedure against beings" (Heidegger 1994, 155). In this there is a risk of arbitrariness, of violently imposing a merely subjective order on things rather than disclosing them in their truth.[4] This notion of arbitrariness can be interpreted in two different ways. First, the arbitrary may manifest itself as error on the terms of the culture, deviation from the essential pattern at which production should aim. Second, culture itself may be conceived as arbitrary. Affirming arbitrariness in this second sense involves relativizing any and all meanings. Heidegger's Greeks were fully aware only of the first form of arbitrariness. The later Heidegger argued that modernity is based on the second form, which now prevails as the "technological revealing."

Heidegger contrasts this Greek understanding of making with our modern technology. Technology too is a mode of revealing, but it does not reveal things in their essential nature. Instead what is revealed is a world of resources and components. The meaning of modern artifacts is simply their functional connection to other artifacts in a system of production and consumption. Heidegger calls this system the "enframing" of being.

In the technological revealing, no essences are uncovered. The place of meaning is now taken by the plan and so reduced to human intentions. This may be described as hubristic, although as the technological system gathers momentum it humbles its human creators by incorporating them into its apparatus. Humans too become mechanical parts in systems that surpass them and assign them their function (Heidegger 1977). They begin to interpret themselves as a special type of machine. The proliferation of operating manuals for every aspect of human life from childrearing to divorce to career choices testifies to the enframing of the human. The role of humans in the revealing of being is occluded. We no longer wonder at the meaningfulness of things. The system appears autonomous and unstoppable. This critique of technology was not explicit in Heidegger's writings until the mid-1930s, but he assumed its main points already.

Heidegger's critique does not address any particular technology. Its object is the technological revealing that stems from the modern ambition to dominate all of being. Heidegger argues that this technological impulse is prior to science, by which he means that viewing the world as an object of domination is a condition for understanding it in modern scientific terms. Why? Because technological thinking eliminates the essences that preceded modern science and reduces meaning to function. New cognitive paths are opened when the making of artifacts is so reduced and differentiated from other dimensions of the culture. With the elimination of teleology and ritual significance, nature is available for analysis and quantification, and a modern mathematical and experimental science is finally possible.

Heidegger is most convincing in arguing that knowledge is ultimately rooted in the enactment of meanings in everyday practice. This phenomenological argument against the neutrality and autonomy of knowledge is echoed in contemporary epistemology and sociology. The notion that meanings are to be found not primarily in the mind as conceptual maps but in action as the guiding principles of practical behavior is especially suggestive.[5]

But there is a puzzling risk of self-referential contradiction in Heidegger's approach. Only in the age of technology is it possible to adopt a synoptic view of the history of being such as Heidegger's. What is so special about this epoch? The customary answer is the one given in the first part of this chapter: we moderns know how to differentiate culture from nature. There is a world "in itself" that can be known in its truth scientifically but that is grasped in one or another arbitrary way in a collective subjectivity, a culture. But this view merely recapitulates a technological understanding of being as raw materials subject to a plan. In sum, we can articulate a general theory of the local origin of culture, a "history of being," because we are situated in a culture that understands all meanings as reducible to the subject.

Heidegger must have been aware of the reflexive paradox implicit in this position. No doubt this is why he rejected the cultural terms on which his thought becomes accessible and insisted on his own ontological language. But the historical problem is not solved by a change in language: one still needs an answer to the question of Heidegger's privileged standpoint. He attempted to overcome the paradox in quasi-Hegelian terms, the "owl of Minerva" rising at the dusk of modernity. A "new beginning" places Germany in touch with a new order of meaning that enables Heidegger to think the limitation of modernity as a specific culture.

What could the new source of meaning be? Surely not the arrogant strutting of those "Aryan worthies" intoxicated by newspapers and beer whom Nietzsche had already denounced fifty years before (Nietzsche 1956, 294–295)! Heidegger's essentially dogmatic claim that a new era had begun is untenable, easily refuted by the very modern thought he hoped to transcend. The breakdown of this whole construction led the later Heidegger to a new position based on poetic thinking. But that new position cancels the original reformist intent of his early philosophy.

Although his criticism of technoscience is harsh, Heidegger does not propose a return to the Greek worldview. He recognizes the validity of modern science but challenges its forgetfulness of another order of truth: the truth of revealing. But if regression is not the solution, is there another way to get beyond the technological era? An active attempt to do so, Heidegger claims, would be just more of the same, more technology. He hints at the possibility of renewing the power of art to transform the world and suggests that the very extremity of the disaster into which technology is leading us might inspire a change. His late call for a "free relation" to technology may not imply total resignation, but it is certainly not a program of technological reform. In his last interview he seems to despair, saying, "Only a god can save us" (Heidegger, 1993a).

Marcuse's New *Technē*

I want to turn now to a consideration of technological reform as Marcuse conceives it. Studying his thought helps to see what is wrong with contemporary Heidegger scholarship that struggles heroically with the texts of the master in the interests of some sort of left-wing politics. This is no doubt a minority view, but it has interesting advocates. The influence of Derrida and Foucault is important in this connection, as is the plausible analogy between Heideggerian *Gelassenheit* and some sort of environmental philosophy (Schürmann 1990; Foltz 1995). Environmentalist, anarchist, and postmodern interpretations are offered on this basis.

But the improbability of all these interpretations is clear from Heidegger's last interview, in which he dismissed democracy and praised the Nazi revolution, which, he still claimed, confronted the real problems but in too limited a manner to solve them (Heidegger 1993a, 104, 111). If there is something of value in Heidegger, as I believe there is, it can be extracted only by sacrificing fidelity to his doctrine. The way to get at this worthwhile contribution is critically, not just exegetically.

This is precisely what Marcuse did during his years as Heidegger's assistant. To some extent the influence of Heidegger continued in Marcuse's later thought as well. In what follows I will try to outline the transformation that Heidegger's argument underwent in Marcuse's writings. This cannot be a straightforward procedure since Marcuse reacted so strongly against Heidegger that he substituted similar ideas from other sources for those of his teacher. Heidegger's influence survives as a kind of archeological stratum underneath these later sources, only occasionally emerging into view.

What was it in *Being and Time* that inspired Marcuse to return to the university as Heidegger's student? He later explained that it was the promise of a "concrete" philosophy (Olafson 2007, 116). This promise accompanied the rebellion against scientism that took an original turn in the early twentieth century. Where nineteenth-century romantics protested against reason in the name of passion, phenomenologists developed an analytic of first-person experience that they interpreted as the foundation of the abstractions in which science consists. Founding these abstractions in experience implied a limit to their range of significance and granted experience rather than nature the ontologically fundamental role. For many philosophers, phenomenology was the essential methodological innovation that enabled the turn to a concrete ontology. It was this turn that attracted Marcuse.

What was unusual about Marcuse's situation was his strong political sympathies. He was a revolutionary socialist bereft of party and hope after the failure of the 1919 German revolution. He could not fall back on the reformist Social Democratic Party since it had crushed the revolution and adopted the prevalent scientism as its philosophy.

There were many different diagnoses of the sickness of German socialism, but the one that appealed to Marcuse was laid out most persuasively in 1923 in Georg Lukács's famous book, *History and Class Consciousness*. There Lukács introduced the concept of reification to broaden Marx's original critique of market rationality into a more radical critique of scientific-technical rationality as the dominant cultural form in modern capitalist society. Lukácsian reification involves an objectivistic misunderstanding of the social world as composed of law-governed things subject to theoretical representation and technical manipulation.

Lukács notes the similarity between scientific knowledge and the laws of the market Marx criticized. The market is a "second nature" with laws as pitiless and mathematically precise as those of the cosmos. Lukács writes, "What is important is to recognize clearly that all human relations (viewed as the objects of social activity) assume increasingly the objective forms of the abstract elements of the conceptual systems of natural science and of the abstract substrata of the laws of nature." Like the worker confronted by the machine, the agent in a market society can only manipulate these laws to advantage, not change them. "Man . . . is a mechanical part incorporated into a mechanical system. He finds it already pre-existing and self-sufficient, it functions independently of him and he has to conform to its laws whether he likes it or not" (Lukács 1971, 89).

We are not far here from Heidegger's later critique of technology as a universal mode of thought and action in modernity. But unlike Heidegger, Lukács envisaged a politics of dereification. As a Marxist he argued that the human reality underlying the reified forms can reassert itself and transform the society (Feenberg 2005, chap. 4). Similarly, Marcuse holds out the promise of radical transformation through political action. But Marcuse also takes over much of Heidegger's analysis of ancient Greek thought. Although he does not employ Heidegger's terminology, he has a similar view of the role of meaning in defining a world. And he seems to be in implicit agreement with Heidegger that the Greek idea of making was based on a specific notion of meaning as essence.[6]

Both Heidegger and Marcuse argue that the normative dimension of *technē* is eclipsed in modern technology. In his early courses Heidegger explained that the knowledge associated with production does not merely concern means but more fundamentally the rightful outcome of productive activity. That outcome, the *ergon* or finished work, is present in the means and directs them toward the realization of an *eidos* or essence. Unlike modern technology, *technē* is not value-neutral knowledge but transcends the opposition of ought and is. This contrast returned after the war in Heidegger's "Question Concerning Technology."

It seems likely that Marcuse's understanding of technology was influenced by these concepts, and in fact there are several positive references to this aspect of Heidegger's thought in Marcuse's later work (e.g., Marcuse, 1964: 153–154; Marcuse 1989, 123). There is, however, a significant difference in emphasis in their later work. Whereas Heidegger emphasizes the ritual aspect of essence, Marcuse identifies essence with potentiality. Under the influence of Hegel, Marcuse explained the concept of essence as the highest realization of what appears imperfectly in the world. When Aristotle claims that "Man is a rational animal," he defines what a human being can be at his or her best, not the common condition. Thus essences are in some sense ideals but not for that matter merely subjective.

In this version of the Greek worldview, being has two dimensions, a first empirical dimension, the objects as they are given in experience, and a second essential dimension of ideal form. The tension between the two dimensions is a permanent feature of existence. Things exist and develop in time, striving toward their essential nature. Our understanding of that striving depends on the imaginative grasp of what things can become. It cannot be limited to empirical observation of what they already are. The concept of truth thus applies not just to propositions but also to things, which can be more or less true to their essential nature.

Marcuse argued that the Greeks misread essential tendencies naïvely in terms of the culturally relative assumptions of their time. This set limits to the understanding of the potentialities of women and slaves we can easily transcend. However, the idea of potentiality survives the discovery of these limitations and is still vital to understanding the modern world. Without it there can be no critical reason.

The modern discovery of the constructive power of the subject stands in the way of a return to an uncritical relation to culture, at least in philosophy if not in everyday life. This constructive power is now exercised not only in the spiritual domain of ideas and beliefs but also materially, through technology, which transforms the environment according to human plans and purposes. Modern society dismisses the essences of antiquity as obstacles to the free exercise of human powers. Technical means are stripped of any relation to an objective "truth" of the object they create. The new norms under which technology stands are reduced to the formal requirements of domination.

This formulation recapitulates in a socially concrete form the basic point of Heidegger's critique of technology, that is, the radical de-worlding accomplished by modernity, which shows up in the reification of society to which the individuals are called to submit. The new conformism consists not in obedience to a leader or to customs but more insidiously in submission to the "facts of life" interpreted one-dimensionally as the only possible organization of a modern society. In so adapting, the individuals fall into the worship of the given.

By contrast to the Greek conception, technological rationality reduces everything to a single dimension. The higher world of essences collapses into everyday existence. According to Marcuse, one-dimensionality characterizes modern societies increasingly as they advance. Scientism leads to a rejection of the imaginative relationship to reality in which essential truth is discovered. Without a transcendent reference, the existing society becomes the horizon of all possible progress. The tensions between the two dimensions are redefined as technical problems to which solutions are available on the terms of the given system. Democracy, for example, is defined by the existing institutions and is not held up as an ideal against which to measure them in view of improving them. The one-dimensional society resembles Heidegger's enframed world insofar as it appears as a closed system of technical action that excludes any fundamental change from within.

According to Marcuse, this system has its origins in capitalism. Capitalist enterprise blocks the autonomous development of its human and natural materials in order to extract the maximum profit. The system that evolves out of these origins is essentially alienated, whether it takes a capitalist or communist form. It is a system of technocratic domination that manipulates the underlying population ruthlessly through propaganda and consumerism. As they are absorbed into the large-scale organizations that run a modern society, the individuals' very survival depends more and more on unthinking conformism.

Having posed the problems in this social context, Marcuse believes he can find solutions that were closed to Heidegger. His emphasis on the complicity of technoscience with capitalism suggests the possibility of radical change under a different economic system. Socialism could restore the second dimension.

Marcuse developed this argument as a historical account of the destiny of reason. This account was shared by other members of the Frankfurt School, although he alone proposed a positive alternative. In Horkheimer the equivalent of Heideggerian *technē* is called "objective reason," a reason that incorporates substantive goals (Horkheimer 1947). The origin of reason in the practical necessities of life is clear in this original objective form. Marcuse could thus argue that from the very beginning reason was rooted in a value judgment, a preference for life over death (Marcuse 1964, 220).[7] Modern scientific-technical rationality, Horkheimer's "subjective reason," is a reduction of the earlier form of rationality. When substantive goals are removed from the structure of rationality, only means are left: reason becomes instrumental.

This transformation of reason is reflected in the methodology of the sciences and eventually of all the academic disciplines. Reality is analyzed exclusively under those empirical aspects that expose it to calculation and control. The teleological concept of essence is expelled from science; nature is revealed as an object of technology, and along with it human beings too are incorporated into a smoothly functioning social machine. This is the basis of the world Heidegger hoped to reform with his new beginning. Marcuse looked forward instead to a return of the "objective" dimension of reason in a future socialist society.

Where Heidegger withdrew from history after his disappointment with Hitler, Marcuse persisted in attempting to rethink the socialist alternative in philosophical terms. Humane goals must once again be intrinsic to reason, if not in the form of ancient essences in some new form appropriate to the modern age. These goals cannot be merely subjective but must be disclosed to the subject in the sense that they must have a validating ground that a reason shaped by modernity can recognize and accept. We appear to have returned to Heidegger's problematic of 1933 in search of a better solution than a *Führer*. Did Marcuse find that solution?

What he proposed was the reconstruction of the technical base of society. He argued that this is the key to restoring the unity of ends and means in a modern context. This would be the equivalent of the creation of a modern *technē*, and in fact Marcuse argued that the link between art and craft in antiquity can be restored in a new form. A technology can be devised that pursues idealizing strategies similar to those of art. Misery, injustice, suffering, and disorder shall not be just stripped out of the artistic image of the beautiful but removed practically from existence by appropriate technological solutions to human problems.

This is the notion of the rupture with the continuum of domination, the qualitative difference of socialism as a new form and way of life, not only rational development of the productive forces, but also the redirection of progress toward the ending of the competitive struggle for existence, not only abolition of poverty and toil, but also reconstruction of the social and natural environment as a peaceful, beautiful universe: *total transvaluation of values, transformation of needs and goals.* This implies *still another change in the concept of revolution,* a break with the continuity of the technical apparatus of productivity which, for Marx, would extend (freed from capitalist abuse) to the socialist society. Such *"technological" continuity* would *constitute a fateful link between capitalism and socialism,* because this apparatus has, in its very structure and scope, become an apparatus of control and domination. *Cutting this link* would mean, not to regress in the technical progress, but to reconstruct the technical apparatus in accordance with the needs of free men. (Marcuse 1970, 280)

But is Marcuse out of the woods with these proposals? Not quite. His argument depends on the notion that the values of peace, beauty, and fulfillment he advocates are not simply normatively preferable to their opposites but that their normativity has a basis of some sort in reality. Like the lump of clay the potter must transform in realizing its potential as a pot, a world characterized by violence, ugliness, and oppression falls short of its essential nature. But this approach requires a notion of privation to which a rational *technē* would respond with appropriate remedies, and that in turn implies an ontology Marcuse did not develop. Scientific naturalism is not suited for this purpose, nor is it plausible to return to Aristotle. The alternative at which Marcuse hinted was a phenomenology of aesthetic experience in a very broad sense. But although there are indications of how he might have developed such an alternative, he did not work it out in sufficient depth and detail to successfully challenge the pessimism of Adorno and Heidegger.

Instead Marcuse turned to a rather formalistic argument that relied on the existential validity of a new aesthetic sensibility for at least some marginal groups. The basis of this new sensibility, he believed, was an immanent critique of the society, contrasting its ideals and its achievements. As Marcuse pointed out, this contrast grows ever more scandalous as the rising productivity of technology removes the material alibis for poverty, discrimination, and war.

This argument then grounded the new *technē* in a rational judgment able to supply the criteria of a "transcendent project," a progressive development beyond the existing society. The criteria include technical feasibility at the given level of knowledge and technology and moral desirability in terms of the preservation and enhancement of human freedom and happiness. Furthermore, the transcendent project's rationality would have to be demonstrated through a persuasive analysis and critique of the existing society (Marcuse 1964, 220).

Aesthetic Technology

Looking back now from the perspective of the new century, Marcuse's general position remains convincing primarily as analysis and critique. *One-dimensional Man* is unsurpassed despite a generation of efforts to elaborate Habermasian critical theory and philosophies of "difference" on the basis of poststructuralism and Adorno. The retreat from the concrete that these alternatives represent is distressingly reminiscent of the false promise of concreteness in Heidegger's work.

What has proven fatal to Marcuse's reputation is his hopeful argument for radical social and technical transformation. Yet this aspect of his work is relevant in a new period of crisis and protest largely focused around technical issues such as environmental pollution, energy politics, and the globalization of industry and disease. In the remainder of this chapter, therefore, I will consider some starting points for continuing the general line of argument that Marcuse developed under the contradictory influences of Heidegger, Marxism, and the New Left.

Heidegger and Marcuse argued that the understanding of beings in general, what we would normally call "culture," is rooted in the instrumental relation to reality. That relation evolves historically and in its latest incarnation takes on a particularly destructive aspect. The danger is not merely physical but concerns the substitution of technological rationality for every other type of thought. The subject in a "one-dimensional society" understands neither its own essential involvement in its world nor the potentialities with which that world is fraught.

From within technological culture it seems that all that has been lost in the disenchantment of the world is arbitrary prejudices and myths. According to this view, modern science supplies all the truth that human beings can possibly require. The lifeworld is a poor source of knowledge until its givens have been refined to remove illusory subjective elements. The pursuit of technical efficiency replaces an understanding of the structure of meaning in which experienced worlds consist. The exclusive focus on the means lops off humanly significant dimensions of experience that appear functionally irrelevant.

Both Heidegger and Marcuse were tempted at least rhetorically to accept such a reductionist vision as accomplished fact while giving it a dystopian twist: the triumph of Brave New World. Yet ultimately neither believed the experience could be wholly disenchanted. Heidegger claimed that behind the functional appearances of modernity there lies a mysterious new meaning that is still hidden to us but that may someday be revealed. Marcuse concluded that the very meaninglessness of modern technology situates it within the project of a ruling class. The destruction of all traditional meaning, which is the condition of capitalist technical and economic advance, is simply the other side of the coin of the reinterpretation of meaning in the degraded form of consumerist ideology.

In his later work, Marcuse argued that socialism would have to transform not just the cultural, economic, and political orders but also the underlying technology, indifferent to nature, human life, and the development of human capacities. He did not share Heidegger's belief that the relationship to technology could be "free" independent of its design. Examples of destructive technologies that Marcuse cites include the assembly line, the mass media, and weaponry. If these technologies remain at the core of modern life, no change in our relation to them can save us. But Marcuse could only hint very generally at what the new technology would be like.

Because both thinkers faced a world in which no alternative appeared at the technical level proper, they sought sources of resistance in other domains such as Nazi politics or New Left protest. But they did not adequately explain or justify this departure from the ontologically fundamental role that technical practice holds in their own philosophies. They ended up with such unsatisfactory conclusions because they could find no way to return to the realm of everyday technical experience to discover there the enactment of new meanings that appeal to a modern ground while pointing beyond the current limitations of modern societies. If we can find a closer connection between politics and technology, a more convincing alternative may appear.

Marcuse at least projects a technical solution to the modern conundrum. He calls for a reunification of differentiated cultural spheres in a reformed scientific-technical rationality. Technology, aesthetics, and ethics must be brought together once again in a unified culture. He is especially concerned with the split between science and art. Art extracts possible ideals from the real and so conserves hopes denied by scarcity and oppression. The imagination brings the second dimension to life in art. Here is how Marcuse explained his position in an important unpublished essay:

Only if the vast capabilities of science and technology, of the scientific and artistic imagination direct the construction of a sensuous environment, only if the work world loses its alienating features and becomes a world of human relationships, only if productivity becomes creativity, are the roots of domination dried up in the individuals. No return to precapitalist, pre-industrial artisanship, but on the contrary, perfection of the new mutilated and distorted science and technology in the formation of the object world in accordance with "the laws of beauty." And "beauty" here defines an ontological condition—not of an *oeuvre d'art* isolated from real existence . . . but that harmony between man and his world which would shape the form of society. (Marcuse 2001, 138–139)

This is an astonishing paragraph. Astonishing for its wild utopianism and its total indifference to mainstream academic opinion and especially to Anglo-American philosophical orthodoxies. It is also a profoundly attractive set of propositions for those seeking a radical civilizational alternative to the existing society. But attractive does not necessarily mean convincing. Marcuse could count on a sympathetic audience for such ideas in 1970 when he wrote this text and others like it. We are reading this passage too late—it has been thirty-eight years as I write this—long after the excitement of the New Left has died. Today speculations such as these resonate with our nostalgia rather than our beliefs. But In his afterword to Marcuse's *Towards a Critical Theory of Society*, Habermas warns us not to be smug, situated as we are in the always superior future. He asks us to "do justice to the truth content of Marcuse's analyses" (Marcuse 2001, 237). He is referring to Marcuse's critique of advanced industrial society, but I believe that the same approach to the positive idea of a redeemed science and technology is also worth attempting.

It is not an easy task to interpret Marcuse's views in a way that communicates directly with us today. As Habermas notes, Marcuse presented the "truth content" of his analyses in "concepts that have become foreign to us" (Marcuse 2001, 237). I will try here to reformulate some of his insights at the risk of modifying them. Let me begin, however, by simply elucidating the meaning of the text I have cited on its own terms.

The notion that a new technology could follow the "laws of beauty" is a direct quotation from Marx's 1844 *Manuscripts*. Marx claims there that while the animals appropriate nature only to satisfy their needs, "man constructs also in accordance with the laws of beauty" (Marx 1963, 128). Marcuse draws on Freud's theory of the erotic to develop this brief mention of beauty in Marx. He argues that the erotic impulse is directed toward the preservation and furtherance of life. It is not merely an instinct or drive but operates in the sensuous encounter with the world. But this impulse is repressed by society, partially sublimated, partially confined to sexuality. The loss of immediate sensory access to the beautiful gives rise to art as a specialized enclave in which we perceive the trace of erotic life affirmation.

The concept of the "aesthetic" is ambiguous, as Marcuse points out, "pertaining to the senses and pertaining to art" (Marcuse 2001, 132). This ambiguity is not merely semantic but stems from a common structure. Marx claimed that the senses "become directly theoreticians in practice" (Marx 1963, 160). They are thus not passive, as empiricism would have it, but engage actively with their objects. Like the practice of art making, the "practice" of sensation involves on the one hand objects rich in meaning and on the other hand subjects capable of receiving that meaning. There is a hierarchy of sensation, going from a minimal, crude encounter with the object through the full realization of its complexity. A dog may hear a symphony, but it will not hear what its master hears. The content of experience is gradually revealed as civilization advances. The human being at home in the world under socialism will find more in nature than does the impoverished and alienated worker under capitalism.[8]

According to Marcuse, aesthetic form is a kind of reduction and idealization that reveals the true essences of things sensuously, things as they would be redeemed in a better world. Form is active in sensation as well, giving rise not only to appreciation of beauty but also to a critical repulsion toward all that is life destroying and ugly. Marcuse argued that the New Left and the counterculture gave us a hint of what an aestheticized sensorium would be like.

Art and technology originate in different faculties. Technology is a product of reason while art has its roots in the erotic imagination. In the past, reason aimed not just at the empirically given but also at discursive comprehension of the ideal form of its objects, their essence, that is, the objects' fulfilled potentialities as conceived by the imagination. Thus art and reason are not entirely alien to each other since they each reveal essences in their own way (Marcuse 1964, 220, 228). But they have been separated by the pressures of life in class society. While art has been confined to a marginal realm of "affirmative culture," reason has been reduced to an instrument in the struggle against scarcity. This remnant is what we mistakenly take for the true nature of rationality.

Corresponding to this reduction of reason, technology is reduced to a value-free means serving functional goals. But value freedom is simply a tendentious way of signifying the differentiation of technology from ethics and aesthetics that restricted it to culturally secured designs and goals in premodern societies. So differentiated, technology is available for any use whatsoever. Ends now come from the users and are subjective. This seems to mean that modern science and technology are innocent of their most terrible applications.

But Marcuse argues that they appear innocent only when artificially separated from their social context. In that larger context, the means they supply are bound up with the practice and the goals of the dominant social subject. Concretely, value neutrality means the overthrow of all restraints on power. Thus,

it is precisely its neutral character which relates objectivity to a specific historical Subject. . . . Theoretical reason, remaining pure and neutral, entered into the service of practical reason. The merger proved beneficial to both. Today, domination perpetuates and extends itself not only through technology but *as* technology, and the latter provides the great legitimation of the expanding political power, which absorbs all sphere of culture. (Marcuse 1964, 156, 158)

The apparent paradox of conflating value freedom and domination dissolves in Marcuse's two-dimensional ontology. Neutrality as between the developmental potential of objects and arbitrary goals is not truly neutral. A rationality that cannot distinguish between the essential growth and development of human and natural beings and such narrow purposes as military power or profit lends itself to the capitalist project of domination. So-called neutral reason is in fact destined to serve those with the power to use it for their arbitrary ends. Its form is appropriate to their needs. In this sense its apparent neutrality is in fact a bias toward domination (Marcuse 1964, 132, 146–148).

The elements are now in place for a radical revision of the concept of technology. Marcuse projects a possible future in which the life-affirming *telos* of art and reason would come together under the aegis of an eroticized sensuousness. The result would be a transformation of technology and therefore of the life environment, which is increasingly mediated by technology. Different human beings would inhabit this world with different perceptions and concerns. This would be a socialism that changed not merely some superficial political and economic formations but the structures of reason, art, technology, and experience itself.

To sum up, the art and technology of the existing society deviate from original forms that were richer and more unified. Art and technology once merged in practices directed toward the realization of the highest forms of their objects, essences, beauty. Experience and reason were once informed by imagination and sensitive to the erotic impulse that joined them in the appreciation of the essential in things. The emancipatory potential of these original forms could not be realized under premodern conditions of scarcity. Today a socialist revolution can revive them and fulfill that potential in rich modern societies.

Science, Technology, and Lifeworld

This summary should make clear what Habermas meant when he claimed that Marcuse's "concepts . . . have become foreign to us." We find it difficult to accept an argument based on a notion of origins such as this one.

The Freudian reference is also less convincing today than it used to be. Furthermore, in Freud the erotic is a subjective drive rooted in human physiology, whereas in Marcuse it uncovers objective structures of being. Marcuse's teleological notion of reason also presupposes a similar ontologizing of anthropological categories. Life affirmation is an existential category and not simply an instinctual drive. Thus a reason that incorporates the affirmation of life in its structure is in harmony with the nature of things in a way that value-neutral reason is not.

For these ideas to make sense, the concept of essence must be reconstructed and revived. The empirical form of human beings and things cannot be the last word on their nature. They are haunted by a negativity that refers us to their potentialities. The erotic, the imagination, the affirmation of life all point to dimensions of being that transcend the given. Hegel enables Marcuse to interpret the concept of essence in a modern vein as the potentialities revealed in the historical process. Marcuse believes we have reached a stage in that process when the gap between existence and essence can be closed by a new technology responsive to values.

Strange as all this sounds, its elements taken one by one are not entirely alien to phenomenological trends that still represent an influential alternative to naturalism and Kantianism. The key missing element in Marcuse's presentation of these ideas is the phenomenological notion of "lifeworld." Although he mentions something he calls an "aesthetic *Lebenswelt*" on several occasions, he never elaborates its phenomenological background (Marcuse 1969, 31). This background is helpful in reconstructing Marcuse's redemptive vision.

The key problem is the ontological status of lived experience. The nature of natural science is totally disenchanted. It has no room for teleology, for the erotic, for any preference for life over death. Like Melville's white whale, it is bleached of value and so invites subjective projections of every sort in the form of ever-more-powerful technologies serving ever-more-violent ends. Against this background, experience is devalued in modern times relative to the scientific picture of nature.

Marcuse rejects the privilege of nature in this scientific sense. Experience is not a subjective overlay on the nature of natural science. It reveals dimensions of reality that science cannot apprehend in its present form. These dimensions, beauty, potentialities, essences, life as a value, are just as real as electrons and tectonic plates. The imagination that projects these dimensions is thus not a merely subjective faculty but reveals aspects of the real.

So far, so good. But there is an ambiguity in Marcuse's approach that shows up in his rather vague demand for a new science that would discover value in the very structure of its objects. Without more to go on, we are left suspended between two possible formulations of his program (Marcuse 1964, chap. 9).

Does Marcuse wish to re-enchant the disenchanted nature of physics and biology, to attribute qualities such as beauty to it that they do not recognize today? Presumably these qualities would appear as phenomena for a science of the future. Could this be the meaning of his enigmatic call to recognize the "existential" truth of nature? He goes on to claim that "The emancipation of man involves the recognition of such truth in things, in nature" (Marcuse 1972, 69). Elsewhere he carries this argument unhesitatingly to the startling conclusion that there are "forces in nature that have been distorted and suppressed—forces which could support and enhance the liberation of man" (Marcuse 1969, 66). Marcuse was thinking primarily of the visible beauty of nature, which he saw as symbol and bearer of peace and happiness. But it is difficult to imagine beauty or any other life-affirming "force" as a variable in the equations of physicists. These statements would seem to require a complete break with science as we know it, but Marcuse explicitly rejects any return to a "qualitative physics," that is to say, to a premodern form of knowledge (Marcuse 1964, 166).

It is obvious that Marcuse's notion of nature is intentionally provocative. Although it lends itself to misinterpretation, a charitable reading is possible. Perhaps we can find a less romantic equivalent for his aesthetic ontology in those aspects of the natural world that support life directly and immediately. Some of these are so obvious as to seem trivial—clean air, abundant water, a climate suitable for agriculture and human life—and yet they are being destroyed by uncontrolled development. These benign natural forces were recognized as such and celebrated by primitive peoples. Respect for such forces is still required of us moderns. Marcuse argues that violence against nature reflects the violence of social relations in a repressive society. Environmentalism and Marcuse's critical theory are thus natural allies.

According to this interpretation, experience is revalorized not in opposition to science but as a coexisting ontological field that claims its own rights and significance. Presumably "existential truths" revealed in experience could inspire new directions for scientific research and technological development in a socialist society without replacing current science with a problematic successor.

This second interpretation is more plausible, but it remains to be seen how it differs from a mere change in the *use* of science of the sort that both Heidegger and Marcuse would surely have dismissed as insufficiently critical. It would be disappointing to return after all these complexities to a commonsense position. Indeed, according to their arguments, nothing fundamental would change if organizations still wielded neutral technology in the interests of arbitrary—if life-affirming—goals. Heidegger's pessimism would seem to be confirmed by such a meager upshot of Marcuse's version of the critique.

I do not find a clear resolution of this ambiguity in Marcuse. Rather than *interpreting* him on this difficult point, I will try to *reformulate* his argument in a way that conforms loosely with his intent. I can only sketch this solution briefly here, but I want at least to hint at it to show that the path we have been following with Heidegger and Marcuse is not a dead end.

Both of these thinkers block the obvious solutions of the sort that lead to cultural dogmatism or New Age re-enchantment. They agree that we cannot return to eternal essences of the sort that guided the Greeks. Tradition no longer has this force in modern societies, and in any case culturally established essences would appear to us moderns as arbitrary restrictions on our freedom. Nor can we re-create lost meaning by an effort of will. That would simply make a technology out of culture and reconfirm the technological enframing. A different model is needed that is neither premodern nor modern in the usual sense of the terms.

This third alternative corresponds to the phenomenological approach as it is explained in thinkers such as Gadamer and Merleau-Ponty. They do not endorse a regressive re-enchantment of nature but defend experience against naturalistic reductionism. In the phenomenological concept of the *Lebenswelt*, the lived world, value, and fact are joined as we have seen in the discussion of Heidegger. Our original encounter with nature, both external nature and human nature, is not objectivistic but practical. In everyday experience we always work with "materials" that possess meaning and seek form. This aspect of Heidegger's phenomenology resembles

Marx's insight according to which the senses are "directly theoreticians in practice" and supports Marcuse's two-dimensional ontology of experience.

A similar phenomenological translation can save Marcuse's notion of the erotic, which must be freed from its physiological formulation in Freud's theory to serve Marcuse's purpose. Marcuse's "erotic" resembles Heidegger's concepts of "attunement" or "state-of-mind." These concepts refer to the fact that sensory experience is always colored by a general quality of perception such as fear or anxiety, joy or hope. These qualities reveal the world in its various aspects and are not merely subjective. The erotic appears to do the same work in Marcuse's argument as one among the possible attunements of which human beings are capable. But unlike the Heideggerian equivalents, such as anxiety and boredom, it reveals the negativity of the world against a normative background. Erotically informed perception is sensitive not only to what is life affirming and but also to what is life denying.

In this phenomenological context, it makes sense to claim that the perceived potentialities of objects have a kind of reality. There are important domains of experience to which we bring a normative awareness quite apart from opinions and intellectual constructions. When we encounter a beautiful landscape, we perceive its beauty immediately without forming a discursive judgment. A sick person appears to our perceptions to fall short of the norm of health to which we expect visible conformity. The examples could be multiplied indefinitely. They show that the lived experience of the real is not confined to the empirically given but frequently refers beyond it to essential potentialities it more or less fulfills.

This "two-dimensionality" of experience has political significance for Marcuse. How we see the world conditions our actions. Where strip malls appear as conveniences, they may be acceptable. Where they appear as ugly defacements of the neighborhood, they will be resisted. Torture perceived as a practical necessity is not the same as torture perceived as a hideous assault on the humanity we share. The sight of workers on an assembly line may evoke thoughts of efficiency, or it may reveal the dehumanizing order of an exploitative economic system. In each of these examples, a one-dimensional perception of contemporary realities sanctions them, while a two-dimensional perception contrasts them with potentialities they foreclose. In the absence of Marx's brand of revolutionary class consciousness, Marcuse turns to such a life-affirming sensibility for a new basis for political resistance.

The Complementarity of Nature and Experience

These reformulations of Marcuse's approach raise a further question concerning the relation between the two worlds, the natural world of science and the lifeworld of experience. Marcuse hoped the lifeworld under socialism could give a new direction to science and technology, but he did not explain how this was to come about. Today, after so many struggles around technology Marcuse did not live to see, we can go beyond merely gesturing at an answer to this question.

The experienced lifeworld and the nature of natural science do not just coexist side by side. They interact in many ways. In the first place, science presupposes meaningful human action through which scientific data are gathered. Experiments, which create closed domains within which laws can be observed to operate, themselves depend on such action. But action is understandable as such, that is, as meaningful, only from an experiential standpoint distinct from that of natural science. When action is reduced to its natural conditions, for example, certain muscular reflexes, it is deworlded and no longer makes sense. If only scientific explanations are valid, as a naturalistic reductionism would have it, then action in the usual sense of the word is eliminated and the possibility of scientific understanding itself rendered unintelligible. In this sense, Karl-Otto Apel argues, action is a quasi-transcendental precondition of (scientific) knowledge. In opposition to naturalistic reductionism, Apel posits the "complementarity" of hermeneutic understanding and scientific explanation (Apel 1984, 63–64).

But Apel's argument is incomplete. His thesis according to which meaningful action is a precondition of scientific knowledge depends on the still-more-fundamental thesis that the precondition of action is the world as the network of meaningful objects. For action to make sense, it must address objects that themselves possess significance. The essential structures of action must correlate with essential structures of objects as they are found in lived experience. The soldier's salute is not more meaningful than his uniform. The seminar table speaks of education, just as the automobile signifies the status of its owner, the lock the fact of ownership, the telephone human contact. The lifeworld includes the whole practical realm, not action alone. But this means that it is not only action but also things that escape reduction.

This observation has implications for technology, which, like scientific experiment, exists on both sides of the line separating the lifeworld from the order of natural causality. Technologies are at one and the same time meaningful within the lifeworld and functional as causal mechanisms. Their two-sidedness is essential to their very being and is not an external combination of subjective feelings and objective things. Meaning is thus the precondition not just of the scientific rationality but also of technology's very existence within a lived world.[9]

Marcuse's attempt to unite art and technology in a value-oriented concept of technological rationality finds support in these ideas. The technological implications of his approach could be developed independently of the hope in a new science that appears to commit him to a re-enchantment of nature he does not need to support his political argument for a nonrepressive society.[10] The argument would then claim that technical disciplines could be restructured under the aegis of values such as beauty that are revealed in experience. The arts would appear not as antagonistic to technology but rather as informing it through revealing the potentialities of its objects. Something like Schiller's "aesthetic education" would be at work here, but it would not be confined to character, as it is in Schiller, but would extend to the technological environment in which and through which the individuals live.

Can we make sense today of this vision of an aestheticized technology? Surprisingly, the answer to this question is "Yes." In fact we are better able to develop this idea than we were in Marcuse's day. This is because the traditional notion of technology as a pure rational "means" to subjective "ends" has been decisively refuted by philosophy and sociology of technology. We no longer believe that technology is value neutral. Rather contemporary technology studies argues that technological design always incorporates values through the choices made between the many possible alternatives confronting the designers. Technologies are not mere means but shape an environment in terms of an implicit conception of human life. They are inherently political. But if this is so, Marcuse's argument gains in plausibility.

As we have seen, Marcuse claimed that the problem with modern technology stemmed from its value neutrality, an effect of its differentiation. Although he did not develop a proper historical account, he appears to have believed that premodern technical activity was guided by values incorporated into the standards and practices of craft, values that reflected a wide range of human needs. The stripping away of these constraints on modern technology turns it into an instrument of domination by the powerful.

This critique of value neutrality is not entirely compatible with contemporary views, but it can be reformulated in a way that preserves Marcuse's essential point. Value neutrality is not an achieved state of purity but a tendency with a history. The imperatives of the capitalist market underlie this tendency to free technology from craft values to a development oriented exclusively toward profit. Naturally the pursuit of profit mediates real demands that shape technical disciplines and designs. No complete value neutrality is ever achieved, but valuative constraints on design are increasingly simplified and rendered ephemeral and controllable. The less technology is invested with preestablished values, the more easily it can be adapted to the changing conditions of the market. Hence the appearance of value neutrality of modern production, with its purified technical disciplines to which correspond standardized parts available for combination in many different patterns with different value implications.

Reformulated on these terms, Marcuse's argument leads to the conclusion that technical disciplines and technologies should be constrained by values related not just to profitability but more broadly to human and natural needs recognized in experience and validated in political debate. The situation Marcuse foresaw is anticipated by the regulation of technology where it imposes life-affirming standards independent of the market. Socialism would represent a shift in the balance toward far more extensive regulation based on far more democratic and participatory procedures.

This conclusion requires further specification in the larger context of Marcuse's radical ontological argument. Recall that according to Heidegger, essences are the form and purpose of materials. But form and purpose are precisely what have been replaced by arbitrary plans in modern times. Yet there is a dimension of essence that is not subject to arbitrary manipulation, the *peras*. For the Greeks, essences are limitations on the formless materials from which the produced thing is made. Meaning arises from selection. The artifact is "forge[d] . . . into its boundaries" (Heidegger 1995, 118). What is excluded is the erroneous move that deviates from the essential *eidos* of the produced thing.

For us moderns, who have lost the essential discrimination of the Greeks, another kind of exclusion is possible. Today limits emerge in the lifeworld as threats to human health or nature that feed back into technology, guiding demands for less destructive designs. The discovery of a limit reveals the significance of that which is threatened beyond it. This dialectic of limitation joins the two worlds. On the one side, the experienced world gains a ground in respect for an object, such as the human body or a threatened natural system. On the other side, a concrete technical response based on knowledge of the objective world employs the means at hand in new combinations or invents new ones.

This is the form in which the lived world we have discovered in the thought of Heidegger and Marcuse becomes active in the structure of a rationality that still has for its mission the explanation of objective nature. Even if this world has no scientific status, the normative concepts that shape it, such as human health and the balance of nature, do not contradict the cognitive advances of modern science but on the contrary require scientific knowledge to evaluate conflicting claims. It is here that we encounter the peculiar ingression of objectivity into experience that corresponds to Apel's account of the foundational role of experience in science. The complementarity of objectivity and experience he identifies is not just cognitive but also has political and technological implications.

No return to a qualitative science is possible or necessary. Modern science objectifies and reifies by its very nature, but it could take into account limits standing in for the lost essences of antiquity and like them referring us to an irreducible truth of experience. Although the process character and full complexity of reality cannot be reflected immediately in the scientific-technical disciplines, they can be deployed in fluid combinations that reflect the complexity of reality as it enters experience through humanly provoked threats and disasters of all sorts and through innovations that offer new perspectives on reality. Specialization and differentiation would not disappear, but they would be treated as methodologically useful rather than as ontologically fundamental. The resultant breaching of the boundaries between disciplines and between the technical realm and the lifeworld responds to the crisis of industrial society. We may learn to bound the cosmos in modern forms by attending to the limits that emerge from the interactions of domains touched by powerful modern technologies.

There is a risk of resignation in this conception that is manifested in many calls to voluntary simplicity and technological regression. But limitation in the Greek sense is not just negative; it is implicated in the positive act of production: *telos* is the other side of *peras*. We must exercise a productive restraint leading to a process of transformation, not a passive refusal of a reified system. As design is pulled in different directions by actors attempting to impose their differing requirements, innovations must reconcile multiple functions in simple and elegant structures capable of serving them all. This is what Gilbert Simondon calls "concretization," designs that accommodate a wide range of influences and contextual factors.[11] Examples abound: hybrid engines in automobiles, refrigerants and propellants that do not damage the ozone layer, substitutes for lead in consumer products, and so on. In the process of developing these technologies, environmental, medical, and other concerns are brought to bear on design by new actors excluded from the original technological regime. Of course no small refinements such as these can resolve the environmental crisis, but the fact that they are possible removes the threat of technological regression as a major alibi for doing nothing.

The larger goal is not merely to address particular problems as they arise but to reconstruct modern technology around a new model of wealth that is environmentally compatible and that draws on human capacities suppressed or ignored in the present dispensation. Marcuse interpreted this in terms of the surrealist *"hazard objectif,"* the rather fantastic notion of an aesthetically formed world in which "human faculties and desires . . . appear as part of the objective determinism of nature—coincidence of causality through nature and causality through freedom" (Marcuse 1969, 31).

This reformulation of Marcuse's project recovers some aspects of the traditional concept of essence but not its cultural rigidity. The negative side of essence, the notion of *peras*, is secured by our knowledge of the limits of the human body and nature. This establishes the boundaries within which creative activity must go on. The new limits make sense in modern scientific terms but cannot be derived from science alone. Can these limits take the place formerly occupied by essence as a mediation between experience and rationality? In part. We may determine scientifically what *not* to do in order to save a forest or a coral reef, but science cannot tell us what to do with the resource thus liberated. Nor can tradition inform our decisions. In this we moderns are left on our own. We must decide in terms of our imaginative sensitivity to the requirements of the good life. This is the precondition for freedom and the free development of human beings in history.

Conclusion

In conclusion I would like to summarize the core of the argument the strands of which I have been following throughout this chapter. The concept of essence that prevailed until the scientific revolution gave rational form to the teleological structure of everyday experience. In modern times, the differentiation of scientific-technical rationality from everyday experience split the two formerly interwoven domains into fragments of an unattainable whole. Under this new dispensation, meaning and ends appear subjective, nature and means objective, and no mediation reconciles them. An earlier form of rationality based on a teleological interpretation of experience is irretrievably lost except as a reminder of that impossible reconciliation.

Today we confront a world of artifacts so elaborate and complex that it overshadows our lives in every domain. But this world is not shaped by essences. Its structures correspond to the various disciplines and organizations that make up modern societies. Until recently it was possible to imagine that the fragmented logic of modernity reflected the nature of reality and the conditions of progress. No longer. The environmental crisis that results from the interference between the fragmented domains reveals the complexity of the real world, which does not respect the boundaries between the historically evolved disciplines and organizations.

The problem reduced to its simplest terms is the collapse of any notion of rational ends once essences no longer guide practice toward sanctioned results. But this formulation masks the deeper question of the nature of these essences in which we can no longer believe. In premodern societies, the concept of essence derived from the making of artifacts according to culturally accepted rules. Essences thus joined experience as it was lived in a particular society with technically rational practices. The artifacts themselves faced in both directions, on the one hand participating in the normatively informed world of everyday experience, on the other hand implementing rational understanding of nature. The two sides merged in less differentiated premodern societies. We have articulated technical practices in theoretical knowledge in modern times while eliminating the experiential dimension of artifacts from our theories.

We cannot recover the normativity of technique by a simple act of will. Norms can emerge only from the shared experience of a community with its world. Worlds in this more or less Heideggerian sense must be understood as realms of practice rather than as a passively observed nature to which "values" are ascribed. Worlds are built out of myriad connections uncovered in the course of everyday experience, as Heidegger explains in the suggestive first part of *Being and Time*. These form a horizon within which actions and objects take on meaning. Meanings are not things we have at our disposal, but frameworks, perspectives that we inhabit and that contribute to making us what and who we are. Meanings are enacted in our perceptions and practices. They are not chosen, but rather they "claim us" from "behind our backs" (Simpson 1995, 47). What might be the source of such meanings today?

Marcuse argued that reason itself should play this role. Reason has always presupposed a value judgment, the preference for life over death. In ignoring this value judgment, modern societies become unreasonable in their very rationality. This formulation evokes a rather limited utilitarian framework, but the problem goes much deeper. The elimination of any value judgment from the structure of modern technological rationality, the neutralization of reason, leads to the collapse of the exclusiveness that is a condition for action in the proper sense of the term. The prevailing technological rationality is thus deficient not only in its indifference to life but also, underlying that indifference, in its very structure. Crudely put, when meanings become marketing devices, anything goes, and rationality is threatened. This threat appears in the growing manipulation of science for corporate advantage, which, projected to the limit, signifies the end of science itself (Michaels 2008).

Our growing sense of the danger of the reified institutions and evermore-powerful technologies bequeathed us by several centuries of capitalist progress confronts us with choices in the remaking of the technical world. At the dawn of the modern era, thinkers such as Descartes and Bacon expected that the new science and technology would be framed by a wisdom restraining human ambitions. Like technology, wisdom too is located between reason and experience. These two modes of thought require each other. This was the original vision of the philosophers who overthrew ancient teleology. But they were unable to find a substitute for essence capable of serving in its place. Perhaps now, at a decisive turning point along the road they opened, we will be able to complete their project.

Afterword

Michel Callon

Can technology—often accused, and rightly so, of silently perpetuating the domination of a minority over a majority—contribute to enriching democratic life? To answer this question, raised time and again, Andrew Feenberg interprets such influential authors as Heidegger and Habermas with insight. He enables us to understand why these respected thinkers were wrong, and in which directions their philosophical reflection should either be pursued or revised. In so doing, he shows the possibility of a philosophy of technology that is not limited by a sterile and repetitive criticism of modernity, and that opens onto new theoretical and practical perspectives. Feenberg's perseverance and constant rigor have enabled him to put the philosophy of technology back on the right track. Rid of its false humanist accents, it is actually surprisingly close to Science and Technology Studies (STS). This afterword is a brief discussion on the points of convergence between the two.

There is no one best way for technological development: this is the first result on which there is unquestionably broad agreement between Feenberg's philosophy and the work of STS, as well as that of political science and the economics of innovation. At any point in time a multiplicity of trajectories opens up to the actors. If one of those trajectories ends up prevailing and thus excluding the alternative options that were initially perceived and considered, the reason is likely to be, not its intrinsic qualities, but historically contingent factors. History does matter. If the option finally chosen seems superior to the others (for instance, the combustion engine that made the development of electric vehicles difficult or even impossible, or a technological standard that prevails owing to the network externalities created by the first adopters), it is because it has benefited from incomparably greater technical, scientific, economic, and political investments than have the alternatives. As the economists of innovation say, it is not because it is superior that a technology is chosen, but rather because it is chosen that it becomes superior. This does not mean that all technologies are of equal worth, but simply that no one can say before-hand, with unquestionable certainty, which one is best suited to the situa-tion. To remove this uncertainty it is necessary to invest, and simply the fact of investing in a particular direction destroys the very possibility of comparison, all else being equal. Feenberg expresses these phenomena of path dependence and technology lock-in—now soundly established facts—in an original, elegant, and truly philosophical way with the notions of "layering" and "branching." Layering describes the mechanisms through which cultural codes are permanently embedded in technologies, while branching denotes all the socio-technical virtualities that, at a given point in time, simply need to be actualized. Keeping the future open by refrain-ing from making irrevocable decisions that one could eventually regret, requires vigilance, reflection, and sagacity at all times. Politics, as the art of preserving the possibility of choices and debate on those choices, is therefore at the heart of technological dynamics.

Technological development—and this is the second point of convergence—has always had the effect of triggering the creation of groups that feel affected by its consequences. The diversity of socio-technical configurations that can be actualized at a given point in time depends on the existence and expression of multiple expectations, projects, problems to solve, and claims made and expressed by these groups, which relent-lessly criticize, analyze, and interpret existing technologies to show their limits and undesirable effects. In so doing these groups highlight certain possibilities and identify potential lines of development that until then were overlooked or simply denied. Their analyses and interpretations some-times result in an exploration of solutions and conceivable configura-tions. They may also entail the organization of experiments, tests, and trials designed to assess the realism and advantages of the various alternatives identified. These are obviously dependent on the state of existing technolo-gies, but since they result from the critical action of the concerned groups, they are not automatically determined by them. Whether these groups are concerned about the technologies or, on the contrary, celebrate them, in every instance they engage in a real process of evaluation fed by the problems that they perceive, the projects that they cherish, and the values that they are not prepared to compromise.

The mobilization of the identities formed in this process and in the interpretation that it implies, is constitutive of technology, of what we could call its essence. Feenberg reminds us in chapter 9 that when Michelangelo was asked how he made his statue of David, he replied, "I just cut away everything that wasn't David." Feenberg goes on to say, "Like the statue of David, essences in Heidegger's interpretation of the Greeks are realized not so much through a positive act of production as through the exclusion of the inessential, of that which deviates from the essential nature awaiting realization. *Hence the concept of essence can be thought of as a limit that specifies the thing, the negative of a positive.*" [my emphasis]. To paraphrase Feenberg commenting on Heidegger, we could say that the form taken by technologies stems not only from experts' intervention but also from that of the concerned groups that have been allowed to contribute to their shaping. And just as there are good and bad sculptors who are more or less skilled in seeing what is inessential in the stone they are sculpting, so too there are good and bad ways of identifying and involving (or not involving) concerned groups. To eliminate what is inessential, would the right thing not be—contrary to what we believe—to allow or even encourage and facilitate the intervention of all the groups that consider themselves to be affected and concerned, as early as possible? The specialists' mission would then be to propose blueprints, which would soon be criticized and redesigned. Feenberg's central message in his book is that the essence of things is obtained not by purification but by successive compositions and compromises. The proponents of STS would agree with him on this point.

This analysis—and Feenberg heavily emphasizes this idea—naturally leads to turning technology into a key object for philosophy. The work of interpretation and proposition mentioned above would be pointless, and probably nonexistent, if its origin and application were not in technologies themselves, in their materialities. They make it possible to articulate and implement the different representations, aspirations, and normative demands through which social groups singularize and define themselves: a value that is not embedded in an artifact is an orphan; it rapidly disappears and loses its effectiveness. But there is more: it is by criticizing in its own way the technologies proposed to it that a group starts to exist. It is existing technologies that make it possible to explain these groups' interpretive work, that suggest it and make it visible, explorable, and thinkable; in short, that structure and feed it. This is the precise point on which Feenberg's work is strategic for philosophical thinking: to grasp the essence of a technology it is necessary, with the concerned groups, to plunge to the heart of it and to study its technical characteristics, the alternative options, their evaluations, and the choices that they underlie. Feenberg turns philosophical thinking upside down. Instead of moving away from concrete technologies and going toward universal, ahistorical, abstract, and disembodied definitions of technology, he decides to move closer to the tangible and to consider technologies in their irreducible singularity. Like STS, he puts himself in the exact place in which technologies were designed, tested, criticized, reviewed, and redirected. But he goes even further. Feenberg believes—and on this point STS scholars cannot help but be convinced by his arguments—that technology allows for the coexistence (and even more than that: the coordination) of different worlds, which it makes compatible (and even more than that: complementary). Technology, or rather technologies as a differentiated set of socio-technical, collectively shaped configurations, can be analyzed as systems of translation in action. The belief in the possible existence of a single common world turns out to be. at least potentially, a dangerous illusion, Feenberg tells us. By taking the opposite stance to a certain humanistic philosophical tradition, he adds: everything must be done so that such a common world cannot arise! When they are well designed, technologies rid us of this illusion. In principle, they enable each one of us to disclose and to enact what singularizes an individual and differentiates that person from the others. Technologies—for this is their essence—live off the diversity of interventions, the plurality of cultural codes and interpretative schemas which contribute to making them evolve and to shaping them. Not only do they support this diversity; but by their sheer existence they can ensure that each position, each identity is taken into account by all the others. The extreme case of the French Minitel, concocted by experts and technocrats dreaming of a neutral, efficient, and universal tool for circulating information, witnessed the unexpected and unplanned development of a communication practice that took advantage of proposed technologies to transform them and to imagine other innovations and uses. Thanks to its users, the Minitel thus became an arrangement, an *agencement*, capable of simultaneously producing *both* information *and* communication. It organized a pacific coexistence between the different but interdependent worlds of experts and users. Thus redesigned, the Minitel acted as a common world, without the existence of that world being imposed by anyone at all!

This analysis of technology and its evolution directly leads to a reinterpretation of its links with politics in general and democracy in particular. Science and technology, which we tend to consider as constituting an autonomous sphere of activity, controlled essentially by experts engaged in closed dialogues with political and economic elites, irresistibly enter into the public sphere. These fields become the subject of debates, caught in controversies in which their political, ethical, cultural, economic, and other dimensions are discussed. If public action takes such an interest in technologies, it is not because their content or essence has suddenly changed. Feenberg tells us that any technology, by constitution, calls for a debate stemming from the existence of a process of dual instrumentalization. The first instrumentalization tends to circumscribe interactions, discussions, and opposition to specialists and experts, who focus on the functionalities of the artifacts that they design. The second instrumentalization expands this debate to include groups that consider themselves to be concerned and affected by these technologies, their uses, their effects, and their meaning. These groups endeavor to shape these technologies toward solving the problems caused by the technologies and in support of the values and interpretive norms that the groups themselves promote. By structuring the work of design and social integration of technologies, this dual instrumentalization makes them accessible to non-experts. But, adds Feenberg, there are several ways of organizing the instrumentalizations. Because he is well integrated into the society surrounding him, the craftsman, magnified by Heidegger, is capable both of defining technical functionalities and of accomplishing the social integration of the artifacts that he makes. Modern societies, as described by Habermas, in particular, confine the first instrumentalization to a sphere devoted to it, that of technicians: a sphere whose influence is contained by another sphere, the one that organizes the lifeworld of human beings and where the non-experts, separated from the experts, go about their work of interpretation and integration. Feenberg believes neither in Heidegger's solution (reverting to craftsmen) nor in Habermas' (differentiating society to protect laypersons from technicians); he considers both these solutions to be analytically and theoretically false. Contrary to Heidegger, he maintains that the artisan is not the only figure that successfully integrates these two instrumentalizations. And contrary to Habermas, he reaffirms that they cannot be dissociated. The only solution that respects the essence of technology is the one that makes both instrumentalizations explicit without seeking to confuse or dissociate them. Feenberg upholds Heidegger's idea that technological creation is simultaneously made of decontextualizations (what Heidegger calls de-worlding), which wrench elements from their original world, and recontextualizations (what Heidegger calls disclosure), which rearrange these elements in artifacts so that they recompose one of the new worlds[1]. As noted above, recontextualization calls for the intervention and engagement of the concerned groups. It is the exclusion of these groups—which simply repeats the same process spontaneously carried out by the experts— that leads Heidegger, Habermas, and then Marcuse into a dead end. And it is the fact of taking them into account that, in contrast, leads Feenberg to the only plausible solution: collaboration between the concerned groups and professional specialists; a collaboration which ensures that the dual requirement of decontextualization and recontextualization is met, and simultaneously guarantees that each of them is closely linked to the other. No need to revert to the single figure of the craftsman to maintain the true essence of the technology! No need for separate spheres that impose an artificial rift on technologies that impedes their development! No temptation to summon the mixed codes proposed by Beck! Technical democracy is the only solution that respects the true essence of technology. Every group mobilized endeavors to promote what it considers to be the right technology or, in other words, the right socio-technical configuration; every group engages in an organized work of investigation and experimentation; every group is called on to compromise with the others with a view to finding solutions that are equally satisfactory to all. This work of interrelated creation and evaluation (where everyone asserts individual preferences but also has to take those of others into consideration) implies adequate procedures that the actors have actually contributed to inventing and implementing, in differing ways. Feenberg shows that when technical democracy is accepted, the result is preferable to the one that would have been obtained had the second instrumentalization been dissociated from the first, according to the principle of separation of the spheres, thus extracting technical work from the political debate.

Contrary to the discourses of meritocracy, the essence of technology is democratic. This is revealed in these crisscrossing processes where every choice is an opportunity for experimentation and normative reflection that test the configurations, to establish which ones are both viable and desirable. We can agree with Feenberg that rationality is not absent from this type of approach, which moreover has the advantage of respecting the diversity of expectations and points of view. At no stage is anyone prohibited from criticizing, reinterpreting, and experimenting: reason is at the very heart of the process. And as this activity is never interrupted and pronounces no exclusion, the technologies spawned by the democratic melting pot are temporary and in a sense constantly falsifiable and open to change—giving the democratic machine food for thought. We can therefore see why Feenberg talks of rational democratization, and of democratic rationalization. On all these points I am delighted to see that the analysis proposed by him is close to the one that we have presented in *Acting in an Uncertain World*.

Such a new technological innovation regime does not lead to an impoverishment, a deterioration, an amputation of the technology and of its formidable power to produce unexpected and better worlds. On the contrary, it is entirely oriented toward its enrichment and upgrading. A new perspective is opened: this is no longer a matter of opposing a modernization believed to be synonymous with failure, or of threats to eliminate or risks to control. On the contrary, it is by going deeper into modernization—that is, by acknowledging the creative and moralizing power of technologies when they are shaped democratically—that the difficulties encountered and the injustices, so rightly denounced, can be overcome. Not less modernization, or another modernization, but, allow me the paradox, even more modernization! The best technology is the one in whose design the concerned groups have been represented and have participated.

If we take the philosophical discourse proposed by Feenberg seriously, we have no choice: the democracy so dear to us can survive only by organizing itself around technological innovation; and, conversely, it could be that the rationalization ideal that we cherish has no meaning and future unless it relies on the formidable operator of democracy that technology

might be. No good democracy without technical democracy! And conversely no good technique without democracy! As Feenberg shows, this
democracy, that STS and philosophy could help to establish, favors a
diversity that, in turn, nourishes it. Thus, Tocqueville's somber prognosis
could be invalidated. The worst is never sure: democracy does not inexora-
bly lead to the reign of uniformization and of the tyrannical mediocrity
attending it.

Notes

Preface

1. The implied reference is to the concept of a godlike "view from nowhere." If it were not too cute, one might rephrase the point here as a "do from knowhere," that is, action understood as just as indifferent to its objects as detached knowing.

2. This is the essential insight of the Actor Network Theory of Bruno Latour and Michel Callon. See Callon et al. (2001).

Chapter 1

1. See Langdon Winner's blistering critique of the characteristic limitations of the position, entitled, "Upon Opening the Black Box and Finding It Empty: Social Constructivism and the Philosophy of Technology" (Winner 1991).

2. *Hansard's Debates*, 1844 (Feb. 22–Apr. 22). The quoted passages are found on pp. 1088–1123.

3. The phrase was used by Jean-Paul Sartre in a speech during the 1968 May Events in Paris to describe the effect of the movement.

4. For more on the hermeneutics of technology, see chapter 7.

5. I will return to this question in chapter 8.

6. This example is analyzed in detail in chapter 5.

7. For an approach to social theory based on this notion (called, however, *"doxa"* by the author), see Bourdieu (1977).

8. Recent studies of the "dual nature" of technical artifacts arrive at similar conclusions (Kroes and Meijers 2002).

9. The concept of the bias of technology is further developed in chapter 8.

10. For more on Simondon's concept of concretization, see chapter 4.

11. I will return to a detailed consideration of this theme in the next chapter.

12. The texts by Heidegger discussed here are, in order, *The Question Concerning Technology*, "The Thing," and "Building Dwelling Thinking" in *Poetry, Language, Thought*, A. Hofstadter, trans. (New York: Harper & Row, 1971). I will return to a consideration of Heidegger in the last part of this book.

Chapter 2

1. See, for example, Venkatachalam (2004) and Kopp et al. (1997).

2. See "Girl Worker in Carolina Cotton Mill," http://www.geh.org/ar/letchild/m197701810015_ful.html#topofimage.

3. Stranger still is the notion that, since individual wealth correlates positively with life expectancy, regulations "induce" deaths by reducing disposable income. This "cost" of regulation was brought before the court in a challenge to the Clean Air Act, but the judge was not impressed. For further discussion of the costs of asthma, see the U.S. Environmental Protection Agency's report, "The Benefits and Costs of the Clean Air Act, 1990 to 2010."

4. For more on Commoner's argument for this point, see Commoner (1971) and Feenberg (1999, chap. 3).

5. This is an argument for a culturally informed version of the notion of path dependence (Arthur 1989).

Chapter 3

1. This projection implies the application of a primitive notion of marginal utility under conditions of income equality, which is not, be it noted, a Marxian desideratum. The varying preference for leisure remains as a basis for the rational allocation of labor. Unfortunately, this appears to create a vicious circle: the least popular jobs would have the shortest hours, requiring the recruitment of a large number of workers who would have to be offered still shorter hours at the margin and so on ad infinitum. Still, it is a nice try for 1888!

2. This is what is wrong with the many polemics against Information Age hype. Philosophers fail us when they do not discuss the reality of the technologies they study but merely respond to the silliest prophecies of enthusiasts. As the straw men hit the ground bleeding, we are left wondering what, after all, *is* actually happening. For a more measured approach, see Feenberg and Barney (2004) and the special section on "Critical Theory of Communication Technology" in *The Information Society Journal* 25 (2).

Chapter 4

1. This brief description of the theory gives only a hint of developments described more fully in several of my books (Feenberg 1999; Feenberg 2002). Chapter 8 also contains a more detailed discussion of the theory.

2. For a further development of this concept of world in relation to Heidegger's thought, see chapter 7.

3. For a review of feminist approaches to technology studies, see Wajcman (2004). Critical theory of technology can situate these approaches in the context of a general social critique of rationality (see Glazebrook 2006).

4. For further discussion of concretization, see Feenberg (1999, 216ff) and chapter 9.

5. See, for example, Schivelbusch (1988) and Cowan (1987).

Chapter 5

1. For the concept of *imaginaire technique*, see Flichy (2007).

2. The alternative solution of slow natural growth that built the Internet required far more powerful computers than were available at reasonable cost in the early years of Teletel.

3. Oudshoorn and Pinch (2005).

Chapter 6

1. I formerly called this "expressive design" (Feenberg 1995, 225).

2. For more on the capital goods market, see Rosenberg (1970). Junichi Murata has developed the significance of Rosenberg's analysis for philosophy of technology. See Murata (2002).

3. For an analysis of the debate over Nishida's politics and one of the principal texts under dispute, see Arisaka (1996). For a variety of positions, see Heisig and Maraldo (1995).

4. Miki's place in Japanese Marxism is discussed in Hitoshi (1967).

5. Marcuse makes a similar argument (see chapter 9).

Chapter 7

1. Before I enter into my theme, I should add that I do not intend to survey all the activity in these two very active fields. An overview of the huge literature they have generated is a subject in itself and not my subject here. In particular, I am

leaving out of my account the many scholars who work on concrete problems with a range of tools drawn from both. My justification for this oversight is twofold: first, I have not yet found among these crossovers a satisfactory *theoretical* mediation between the two fields; and second, the most influential figures writing theory in these fields are not seeking such a mediation but on the contrary ignore or exclude each others' contributions. Clearly, this situation deserves treatment on its own terms.

2. The notion of rationality as a cultural form is suggested by Weber's concept of rationalization. Lukács's theory of reification refined that concept by identifying the tensions between the type of rationality characteristic of capitalist society and the lifeworld it enframes. See Feenberg (1986, chap. 3).

3. For explorations of the relation between Marxism and modernity theory, see Berman (1982) and Frisby (1986).

4. There is an enormous literature on Kuhn. For an interesting recent critique, see Fuller (2000).

5. I have reformulated Habermas's position to take technology into account (1999, chap. 7).

6. The early Marxist Lukács already identified this plausible outcome of differentiation, which he called "reification." According to Lukács, capitalist society is characterized by the rationality of the "parts"—individual enterprises, for example—and the irrationality of the whole, leading to recurrent crises (Feenberg 1986, 69–70).

7. I have independently proposed something similar in Feenberg (1992) and Feenberg (1991, 191–198). What I call "subversive" or "democratic rationalization" resembles Beck's "sub-politics," and his "code syntheses" resemble the social interpretation of the theory of concretization. There seems nevertheless to be a difference in our relation to the field of technology studies, which should become clear to readers of Beck in what follows.

8. Richard Feynman defends the standard view of the accident, which he helped to shape. His observations are based not on constructivist methods but on common sense. Feynman's account is devastating for NASA management. Consider, for example, the reaction of programmers to his praise for their very thorough testing programs: "One guy muttered something about higher-ups in NASA wanting to cut back on testing to save money: 'They keep saying we always pass the tests, so what's the use of having so many?'" (Feynman 1988, 194).

9. They reply to my critique in *Studies in the History and Philosophy of Science* 38 (2006). In their reply they do not seem to come to grips with my argument but instead emphasize the unrealistic expectation of reliability with which NASA surrounded the space shuttle. On this point we agree.

10. "[S]i je ne parle pas de 'culture', c'est parce que ce nom est réservé à une seulement des unités découpés par les Occidentaux pour définir l'homme. Or, les forces ne peuvent être partagées en 'humaines' et 'non-humaines', sauf localement et pour renforcer certains réseaux."

11. In his recent book *Reassembling the Social* (Oxford University Press, 2006), Latour attempts to moderate his stance, and he does succeed in making it much more intelligible. But he continues to put forward essentially the views criticized here.

12. For an interesting discussion of the relation of hermeneutics to phenomenological and constructivist science studies, see Egger (2006, chap. 3).

13. Note the similarity between this view and Miki's view as presented in chapter 6.

Chapter 8

1. Those limits show up in periodic crises that reveal the irrationality of the system as a whole. An entirely different kind of irrationality, judged in terms of notions of capacities and freedom, condemns the system for other limits such as the human consequences of factory work.

2. Plato's *Gorgias* contains a much earlier example in Callicles' refutation of civil equality. See Plato (1952, 51).

3. The case offers an interesting parallel to the relationship of sex and gender in Judith Butler's antiessentialist gender theory (Butler 1990). Butler argues that sex does not precede and found gender because our understanding of sex, even in its pure anatomical concreteness, is already shaped by assumptions about gender. I think she would agree that the two are distinguishable in a meaningful way—otherwise there could be no science of sex—but they are not ontologically distinct. Like Latour's hybrids, the body, as a living actor, is ontologically fundamental rather than the two aspects of nature and culture abstracted from it in modern discourses. If there is a problem with this view, it lies in the tendency of its advocates to discount the internally coherent, rational form of the abstractions in which nature is constructed by the sciences.

4. In earlier discussions of the instrumentalization theory, I sometimes included causal relations between devices and between devices and nature under the heading of "systematization." I now realize that this confuses the issue. Every causal relation established at the primary level is paralleled by meanings at the secondary level. These meanings constitute systematizations in the instrumentalization theory. For example, the substitution of a safe alternative for an ozone-depleting refrigerant is a change in the causal relation of refrigeration to nature that depends on a change at the level of meaning signifying that protection of the ozone layer as important. Only this latter change is a systematization on the terms of the instrumentalization theory.

5. Function is also abstracted from a wider range of causal relations, called "effects" in the instrumentalization theory, which includes unintended consequences. See Feenberg (1995, 81).

6. This is an example of meaning as connotation. For the relation of the semiotic concepts of denotation and connotation to the hermeneutics of technology, see Baudrillard (1968).

7. These cases are discussed in chapters 7, 5, and 1 in this book.

Chapter 9

1. See also my account of the relation between the thought of Heidegger and Marcuse in Feenberg (2005).

2. The following exposition is based primarily on *Being and Time* (1962); however, the main lines of that early work are taken for granted by Heidegger until the end, and so this very general description of his thought applies also to the later *Question Concerning Technology* (1977).

3. This notion has its parallel in the derivation of presence-at-hand from readiness-to-hand in *Being and Time*.

4. Heidegger's "reform" of the university was intended to block such arbitrariness by tying scholarship to the limits of a *technē*. At that time, Heidegger considered statesmen to belong to a superior order of producers (see Todorov 2007). The *technē* in question was thus the formation of the Nazi state. The university was to maintain its autonomy precisely through subordinating its understanding of the world to the intrinsic necessities and limits of the national restoration brought about by Hitler. In Heidegger's own mind, this was quite distinct from politicizing *Wissenschaft* by infusing it with political propaganda.

5. As practical enactment, meaning has a "material" dimension that might be explored in a phenomenology of technical practice and technology. This has implications for the discursive turn in contemporary philosophy. So long as reality is understood as structured by or like a language, it is difficult to account for the passive aspect of knowing. The failure to take into account the resistance of the object and the facticity of the subject leads discourse theory to an implausible relativism. But if meanings are understood as enacted in a practice, they cannot be merely subjective but must entertain a relation to a materiality of some sort (Angus 2000, 13). Developing this approach would make sense of the moment of receptivity in such Heideggerian notions as disclosure.

6. The following discussion is based on Marcuse's *One-dimensional Man*, chapters 5–6.

7. Curiously, a similar point is made negatively by Adorno and Horkheimer, who attribute reason to fear of nature, the flip side of Marcuse's positive notion of life affirmation.

8. This theory corresponds to what Adorno refers to as a "mediation" theory of sensation in which both object and subject contribute to the shaping of experience. For an account of Adorno's theory, see O'Connor (2005, chaps. 2–3).

9. This is the import of the instrumentalization theory explained in more detail in chapters 4 and 8.

10. This is not to say that science is unaffected by society. Both science and technology are oriented in their choice of problems by the social environment, and many of their fundamental assumptions depend on the wider cultural background. But it is important to mark a difference in the degree to which the *contents* and *method* of scientific and technical knowledge are vulnerable to direct public influence and government regulation. No doubt this difference lies along a continuum since science and technology are so closely imbricated, but the difference is nevertheless real and politically significant (Feenberg 2002, 170–175).

11. The concept of concretization is discussed further in chapter 4. Important contributions to understanding how the experience of non-experts can improve technical decision making are Collins and Evans (2002) and Wynne (1989).

Afterword

1. In *Acting in an Uncertain World* (Callon et al. 2009), we propose a similar analysis by distinguishing three translations, the first of which corresponds to decontextualization and third to recontextualization.

References

Adorno, Theodor. 2000. *Introduction to sociology.* Trans. E. Jephcott. Cambridge, UK: Polity.

Adorno, Theodor, and Max Horkheimer. 1972. *Dialectic of enlightenment.* Trans. J. Cummings. New York: Herder and Herder.

Akrich, Madeleine. 1992. The de-scription of technical objects. In *Shaping technology/building society: Studies in sociotechnical change,* ed. Wiebe E. Bijker and John Law, 205–224. Cambridge, MA: MIT Press.

Angus, Ian. 2000. *Primal scenes of communication.* Albany: State University of New York Press.

Apel, Karl-Otto. 1984. *Understanding and explanation: A transcendental-pragmatic approach.* Cambridge, MA: MIT Press.

Arisaka, Yoko. 1996. The Nishida enigma. *Monumenta Nipponica* 51 (1): 81–99.

Armstrong, Isobel. 2008. *Victorian glassworlds: Glass culture and the imagination 1830-1880.* New York: Oxford University Press.

Arthur, Brian. 1989. Competing technologies, increasing returns, and lock-in by historical events. *Economic Journal* 99:116–131.

Attali, Jacques, and Yves Stourdzé. 1977. The slow death of monologue in French society, ed. Ithiel de Sola Pool. In *The social impact of the telephone.* Cambridge, MA: MIT Press.

Bakardjieva, Maria. 2009. Subactivism: Lifeworld and politics in the age of the Internet. *The Information Society* 25 (2): 91–104.

Baltz, Claude. 1984. Grétel: Un Nouveau Média de Communication. In *Télématique: Promenades dans les Usages,* ed. Marie Marchand and Claire Ancelin, 184–185. Paris: La Documentation Francaise.

Baudrillard, Jean. 1968. *Le système des objets.* Paris: Gallimard.

Beck, Ulrich. 1992. *Risk society.* London: Sage.

Beck, Ulrich. 1994. *Reflexive modernization: Politics, tradition and aesthetics in the modern social order.* Stanford, CA: Stanford University Press.

Bell, Daniel. 1973. *The coming of post-industrial society.* New York: Basic Books.

Bellamy, Edward. [1888] 1960. *Looking backward: 2000–1887.* New York: Signet.

Beniger, James. 1986. *The control revolution: Technological and economic origins of the information society.* Cambridge, MA: Harvard University Press.

Benjamin, Walter. 1978. Paris, capital of the nineteenth century. In *Reflections,* ed. Peter Demetz, 145–162. Trans. Edmund Jephcott. New York: Harcourt Brace Jovanovich.

Berman, Marshall. 1982. *All that is solid melts into air: The experience of modernity.* New York: Simon & Schuster.

Bertho, Catherine. 1981. *Télégraphes et Téléphones: de Valmy au Microprocesseur.* Paris: Livre de Poche.

Bidou, Catherine, Marc Guillaume, and Véronique Prévost. 1988. *L'ordinaire de la télématique: Offre et usages des services utilitaires grand-public.* Paris: Editions de l'Iris.

Bloor, David. 1991. *Knowledge and social imagery.* Chicago: University of Chicago Press.

Borgmann, Albert. 1984. *Technology and the character of contemporary life.* Chicago: University of Chicago Press.

Borgmann, Albert. 1992. *Crossing the postmodern divide.* Chicago: University of Chicago Press.

Bos, Bram, Peter Koerkamp, and Karin Groenestein. 2003. A novel design approach for livestock housing based on recursive control—with examples to reduce environmental pollution. *Livestock Production Science* 84: 157–170.

Bourdieu, Pierre. 1977. *Outline of a theory of practice.* Trans. R. Nice. Cambridge, UK: Cambridge University Press.

Bowker, Geoffrey, and Susan Leigh Star. 2002. *Sorting things out: Classification and its consequences.* Cambridge, MA: MIT Press.

Branscomb, Anne W. 1988. Videotex: Global progress and comparative policies. *Journal of Communication* 38 (1): 50–59.

Braverman, Harry. 1974. *Labour and monopoly capital.* New York: Monthly Review.

Bruhat, Thierry. 1984. Messageries electroniques: Grétel a Strasbourg et Teletel a Vélizy. In *Télématique: Promenades dans les usages,* ed. Marie Marchand and Claire Ancelin. 51–69. Paris: La Documentation Francaise.

Burke, John. 1972. Bursting boilers and the federal power. In *Technology and culture*, ed. M. Kranzberg and W. Davenport. New York: New American Library.

Butler, Judith. 1990. *Gender trouble*. New York: Routledge.

Callon, Michel, Pierre Lascoumbes, and Yannick Barthe. 2009. *Acting in an uncertain world*. Cambridge, MA: MIT Press.

Cambrosio, Alberto, and Camille Limoges. 1991. Controversies as governing processes in technology assessment. *Technology Analysis and Strategic Management* 3 (4): 377–396.

Certeau, Michel de. 1980. *L'invention du quotidien*. Paris: UGE.

Chabrol, J. L., and Pascal Perin. 1989. *Usages et usagers du vidéotex: Les pratiques domestiques du vidéotex en 1987*. Paris: D.G.T.

Charon, Jean-Marie. 1987. Teletel, de l'interactivité homme/machine a la communication médiatisée. In *Les paradis informationnels*, ed. Marie Marchand, 95–128. Paris: Masson.

Charon, Jean-Marie, and Eddy Cherky. 1983. *Le vidéotex: Un nouveau média local: Enquete sur l'experimentation de Vélizy*. Paris: Centre d'Etude des Mouvements Sociaux.

Collins, H. M., and Robert Evans. 2002. The third wave of science studies: Studies of expertise and experience. *Social Studies of Science* 32 (2): 235–296.

Collins, Harry, and Trevor Pinch. 1998. *The golem at large: What you should know about technology*. Cambridge, UK: Cambridge University Press.

Collins, Harry, and Trevor Pinch. 2007. Who is to blame for the Challenger Explosion. *Studies in the History and Philosophy of Science* 38 (1): 254–255.

Commoner, Barry. 1971. *The closing circle*. New York: Bantam.

Cowan, Ruth Schwartz. 1987. The consumption junction: A proposal for research strategies in the sociology of technology. In *The social construction of technological systems: New directions in the sociology and history of technology*, ed. Wiebe E. Bijker, Thomas P. Hughes, and Trevor J. Pinch, 261–280. Cambridge, MA: MIT Press.

Cunningham, Frank. 1987. *Democratic theory and socialism*. Cambridge, UK: Cambridge University Press.

Dewey, John. 1980. *The public and its problems*. Athens, OH: Swallow Press.

Doppelt, Gerald. 2008. Values in science. In *Routledge companion to philosophy of science*, ed. M. Card and S. Psillo, 302–313. New York: Routledge.

Egger, Martin. 2006. *Science, understanding, and justice*. Chicago: Open Court.

Ehrlich, P., and R. Harriman. 1971. *How to be a survivor*. New York: Ballantine.

Ellul, Jacques. 1964. *The technological society.* Trans. J. Wilkinson. New York: Vintage.

Engels, Frederick. 1970. *The housing question.* Moscow: Progress Publishers.

Epstein, Steven. 1996. *Impure science.* Berkeley and Los Angeles: University of California Press.

Ettema, James. 1989. Interactive electronic text in the United States: Can videotex ever go home again? In *Media use in the information society,* ed. J. C. Salvaggio and J. Bryant, 105–124. Hillsdale, NJ: Lawrence Erlbaum Associates.

Feenberg, Andrew. 1986. *Lukács, Marx, and the sources of critical theory.* New York: Oxford University Press.

Feenberg, Andrew. 1989a. A user's guide to the pragmatics of computer mediated communication. *Semiotica* 75 (3–4): 257–278.

Feenberg, Andrew. 1989b. The written world. In *Mindweave: Communication, computers, and distance education,* ed. A. Kaye and R. Mason, 22–39. Oxford: Pergamon Press.

Feenberg, Andrew. 1991. *Critical theory of technology.* New York: Oxford University Press.

Feenberg, Andrew. 1992. Subversive rationalization: Technology, power and democracy. *Inquiry* 35 (3–4): 301–322.

Feenberg, Andrew. 1993. Building a global network: The WBSI experience. In *Global networks: Computerizing the international community,* ed. L. Harasim, 185–197. Cambridge, MA: MIT Press.

Feenberg, Andrew. 1995. *Alternative modernity: The technical turn in philosophy and social theory.* Berkeley and Los Angeles: University of California Press.

Feenberg, Andrew. 1999. *Questioning technology.* New York: Routledge.

Feenberg, Andrew. 2002. *Transforming technology: A critical theory revisited.* New York: Oxford University Press.

Feenberg, Andrew. 2005. *Heidegger and Marcuse: The catastrophe and redemption of history.* New York: Routledge.

Feenberg, Andrew, and Maria Bakardjieva. 2004. Consumers or citizens? The online community debate. In *Community in the digital age,* ed. A. Feenberg and D. Barney, 1–28. Lanham, MD: Rowman and Littlefield.

Feenberg, Andrew, and Darin Barney. 2004. *Community in the digital age.* Lanham, MD: Rowman and Littlefield.

Feenberg, Andrew, and Jim Freedman. 2001. *When poetry ruled the streets: The French May Events of 1968.* Albany: State University of New York Press.

Feenberg, Andrew, J. Licht, K. Kane, K. Moran, and R. A. Smith. 1996. The on-line patient meeting. *Journal of Neurological Sciences* 139 (Suppl.): 129–131.

Feynman, Richard. 1988. *What do you care what other people think?* New York: W.W. Norton.

Fischer, Claude. 1988a. Touch someone: The telephone industry discovers sociability. *Technology and Culture* 29:32–61.

Fischer, Claude. 1988b. Gender and the residential telephone, 1890–1940: Technologies of sociability. *Sociological Forum* 3 (2): 211–233.

Flichy, Patrice. 2007. *Understanding technological innovation: A socio-technical approach.* Cambridge, MA: MIT Press.

Foltz, Bruce. 1995. *Inhabiting the Earth.* Atlantic Highlands, NJ: Humanities Press.

Forty, Adrian. 1986. *Objects of desire.* New York: Pantheon.

Foucault, Michel. 1975. *Surveiller et punir.* Paris: Gallimard.

Foucault, Michel. 1977. *Discipline and punish.* Trans. A. Sheridan. New York: Pantheon.

Frisby, David. 1986. *Fragments of modernity.* Cambridge, MA: MIT Press.

Fuller, Steve. 2000. *Thomas Kuhn: A philosophical history for our times.* Chicago: University of Chicago Press.

Giraud, Alain. 1984. Une lente emergence. In *Télématique: Promenades dans les usages*, ed. Marie Marchand and Claire Ancelin, 3–10. Paris: La Documentation Francaise.

Glazebrook, Trish. 2006. An ecofeminist response. In *Democratizing technology: Building on Andrew Feenberg's critical theory of technology*, ed. T. Veak, 37–52. Albany: State University of New York Press.

Goffman, Erving. 1982. *Interaction ritual.* New York: Pantheon.

Gottleib, Nanette. 2000. *Word-processing technology in Japan.* Richmond, UK: Curzon Press.

Grimes, Sara M. 2006. Online multiplayer games: A virtual space for intellectual property debates? *New Media & Society* 8 (6): 969–990.

Grimes, Sara, and Andrew Feenberg. 2009. Rationalizing play: A critical theory of digital gaming. *The Information Society* 25 (2): 105–118.

Guillaume, Marc. 1975. *Le capital et son double.* Paris: Presses Universitaires de France.

Guillaume, Marc. 1982. Telespectres. *Traverse* 26: 22–23.

Guillaume, Marc. 1986. *La contagion des passions*. Paris: Plon.

Habermas, Jürgen. 1970. *Toward a rational society*. Trans. J. Shapiro. Boston: Beacon Press.

Habermas, Jürgen. 1984, 1987. *The theory of communicative action: Lifeworld and system: A critique of functionalist reason*. Trans. T. McCarthy. Boston: Beacon.

Habermas, Jürgen. 1986. *Autonomy and solidarity: Interviews*. Ed. P. Dews. London: Verso.

Haraway, Donna. 1991. A cyborg manifesto. In *Simians, cyborgs, and women: The reinvention of nature*, 149–181. New York: Routledge.

Heidegger, Martin. 1977. *The question concerning technology*. Trans. W. Lovett. New York: Harper & Row.

Heidegger, Martin. 1993a. Only a god can save us. Trans. M. Alter and J. Caputo. In *The Heidegger controversy: A critical reader*, ed. Richard Wolin, 91–115 Cambridge, MA: MIT Press.

Heidegger, Martin. 1993b. Overcoming metaphysics. Trans. J. Stambaugh. In *The Heidegger controversy: A critical reader*, ed. Richard Wolin, 67–90. Cambridge, MA: MIT Press.

Heidegger, Martin. 1993c. The self-assertion of the German university. Trans. W. S. Lewis. In *The Heidegger controversy: A critical reader*, ed. Richard Wolin, 29–39. Cambridge, MA: MIT Press.

Heidegger, Martin. 1994. *Basic questions of philosophy*. Trans. R. Rojcewicz and A. Schuwer. Bloomington: Indiana University Press.

Heidegger, Martin. 1995. *Aristotle's metaphysics? 1–3: On the essence and actuality of force* Trans. W. Brogan and P. Warnek. Bloomington: Indiana University Press.

Heidegger, Martin. 1997. *Plato's sophist*. Trans. R. Rojcewicz and A. Schuwer. Bloomington: Indiana University Press.

Heidegger, Martin. 1998. Traditional language and technological language. Trans. W. Gregory. *Journal of Philosophical Research* XXIII:129–145.

Heilbroner, Robert. 1975. *An inquiry into the human prospect*. New York: Norton.

Heisig, John, and John Maraldo, eds. 1995. *Rude awakenings: Zen, the Kyoto School and the question of nationalism*. Honolulu: University of Hawaii Press.

Hitoshi, Imamura. 1967. Marxisme Japonais et Marxisme Occidental. *Actuel Marx* 2:46–47.

Hofstadter, Douglas. 1979. *Gödel, Escher, Bach*. New York: Basic Books.

Horkheimer, Max. 1947. *Eclipse of reason*. New York: Seabury Press.

Huxley, Aldous. 1969. *Brave new world*. New York: Harper & Row.

Iwaasa, Raymond-Stone. 1985. Télématique grand public: l'information ou la communication? Les cas de Grétel et de Compuserve. *Le Bulletin de l'IDATE* 18.

Jouet, J., and P. Flichy. 1991. *European telematics: The emerging economy of words*. Trans. D. Lytel. Amsterdam: Elsevier.

Kirkpatrick, Graeme. 2004. *Critical technology: A social theory of personal computing*. Aldershot, UK: Ashgate.

Kopp, R. J., W. W. Pommereline, and N. Schwarz, eds. 1997. *Determining the value of non-marketed goods: Economic, psychological, and policy relevant aspects of contingent valuation methods*. Boston: Kluwer Academic Publishers.

Kroes, Peter, and Anthonie Meijers. 2002. Reply to critics. *Technē* 6 (2): 34–43.

Kuhn, Thomas. 1962. *The structure of scientific revolutions*. Chicago: University of Chicago Press.

Latour, Bruno. 1984. *Les microbes: Guerre et paix, suivi de irréductions*. Paris: A.M. Métailié.

Latour, Bruno. 1987. *Science in action: How to follow scientists and engineers through society*. Cambridge, MA: Harvard University Press.

Latour, Bruno. 1992. Where are the missing masses? The sociology of a few mundane artifacts. In *Shaping technology/building society: Studies in sociotechnical change*, ed. W. Bijker and J. Law, 225–258. Cambridge, MA: MIT Press.

Latour, Bruno. 1993. *We have never been modern*. Trans. C. Porter. Cambridge, MA: Harvard University Press.

Latour, Bruno. 1994. Les objets ont-ils une histoire? Recontre de Pasteur et de Whitehead dans us bain d'acide lactique. In *L'Effet Whitehead*, ed. I. Stengers, 197–217. Paris: Vrin.

Latour, Bruno. 1999. *Politiques de la nature: Comment faire entrer les sciences en démocratie*. Paris: La Découverte.

Latour, Bruno. 2006. *Reassembling the social*. New York: Oxford University Press.

Law, John. 1989. Technology and heterogeneous engineering: The case of Portuguese expansion. In *The social construction of technological systems*, ed. Wiebe Bijker, Thomas Hughes, and Trevor Pinch, 111–134. Cambridge, MA: MIT Press.

Lee, O.-Young. 1984. *Smaller is better: Japan's mastery of the miniature*. Tokyo: Kodansha.

Lukács, Georg. 1971. *History and class consciousness.* Trans. R. Livingstone. Cambridge, MA: MIT Press.

Lyotard, Jean-Francois. 1979. *La condition postmoderne.* Paris: Editions de Minuit.

Malm, William. 1971. The modern music of Meiji Japan. In *Tradition and modernization in Japanese culture,* ed. Donald Shively, 257–300. Princeton, NJ: Princeton University Press.

de Mandeville, Bernard. 1970. *The fable of the bees.* Baltimore: Penguin.

Marchand, Marie. 1984. Conclusion: Vivre avec le videotex. In *Le Videotex: Contribution aux debats sur la telematique,* ed. Claire Ancelin and Marie Marchand, 175–186. Paris: La Documentation Francaise.

Marchand, Marie. 1987. *La grande aventure du Minitel.* Paris: Larousse.

Marcuse, Herbert. 1964. *One-dimensional man.* Boston: Beacon Press.

Marcuse, Herbert. 1968. Industrialization and capitalism in the work of Max Weber. In *Negations,* 201–226. Trans. J. Shapiro. Boston: Beacon.

Marcuse, Herbert. 1969. *An essay on liberation.* Boston: Beacon.

Marcuse, Herbert. 1970. Re-examination of the concept of revolution. In *All we are saying,* ed. A. Lothstein, 273–282. New York: Capricorn Books.

Marcuse, Herbert. 1972. *Counter-revolution and revolt.* Boston: Beacon.

Marcuse, Herbert. 1989. From ontology to technology: Fundamental tendencies of industrial society. In *Critical theory and society: A reader,* ed. E. Bronner and D. Kellner, 119–127. New York: Routledge.

Marcuse, Herbert. 2001. *Towards a critical theory of society,* ed. D. Kellner. New York: Routledge.

Marx, Karl. 1906 reprint. *Capital.* New York: Modern Library.

Marx, Karl. 1963. *Karl Marx: Early writings.* Ed. T. Bottomore. London: C. A. Watts.

Marx, Karl. 1973. *Grundrisse.* Trans. M. Niclaus. Baltimore: Penguin.

Mauss, Marcel. 1980. Essai sur le don. In *Sociologie et anthropologie.* Paris: PUF.

Mayntz, Renate, and Volker Schneider. 1988. The dynamics of system development in a comparative perspective: Interactive videotex in Germany, France and Britain. In *The development of large technical systems,* ed. Renate Mayntz and Thomas Hughes, 263–298. Boulder, CO: Westview Press.

McCarthy, Thomas. 1991. Complexity and democracy: Or the seducements of systems theory. In *Communicative action,* ed. A. Honneth and H. Joas. Trans. J. Gaines and D. Jones, 119–135. Cambridge, MA: MIT Press.

McCaughey, Martha, and Michael Ayers. 2003. *Cyberactivism: Online activism in theory and practice.* New York: Routledge.

McLuhan, Marshall. 1964. *Understanding media: The extensions of man.* New York: McGraw Hill.

Meadows, D., J. Randers, and W. W. Behrens III. 1971. *The limits to growth.* New York: Universe Books.

Michaels, David. 2008. *Doubt is their product: How industry's assault on science threatens your health.* Oxford, UK: Oxford University Press.

Miki, Kiyoshi. 1967. *Miki Kyoshi Zenshu.* Tokyo: Iwanami.

Milberry, K. 2007. *The wiki way: Anticipating change, practicing democracy.* Wageningen, Netherlands: Tailoring Biotechnologies.

Miller, Richard W. 1984. *Analyzing Marx: Morality, power and history.* Princeton, NJ: Princeton University Press.

Murata, Junichi. 2002. Creativity of technology—An origin of modernity? In *Science and Other Cultures: Diversity in the Philosophy of Science and Technology,* ed. Robert Figueroa and Sandra Harding, 252–266. New York: Routledge.

Murata, Junichi. 2003. Creativity of technology: An origin of modernity? In *Modernity and technology,* ed. Tom Misa, Philip Brey, and Andrew Feenberg, 227–253. Cambridge, MA: MIT Press.

Nickles, Shelley. 2002. "Preserving women": Refrigerator design as social process in the 1930s. *Technology and Culture* 43 (4): 693–727.

Nietzsche, Friedrich. 1956. *The birth of tragedy and the genealogy of morals.* Trans. F. Gollfing. New York: Anchor.

Nishida, Kitaro. 1965a. Sekai shin chitsujo no genri [The principle of new world order]. In *Nishida Kitaro Zenshu,* Vol. 12. Tokyo: Iwanami Shoten.

Nishida, Kitaro. 1965b. Rekishi tetsugaku ni tsuite [On the philosophy of history]. In *Nishida Kitaro Zenshu,* Vol. 12. Tokyo: Iwanami Shoten.

Nishida, Kitaro. 1965c. Nihonbunka no mondai [The problem of Japanese culture]. In *Nishida Kitaro Zenshu,* Vol. 12. Tokyo: Iwanami Shoten.

Nishida, Kitaro. 1970. *Fundamental problems of philosophy.* Trans. D. Dilworth. Tokyo: Sophia University Press.

Nishida, Kitaro. 1991. *La culture Japonaise en question.* Trans. Pierre Lavelle. Paris: Publications Orientalistes de France.

Noble, David. 1984. *Forces of production.* New York: Oxford University Press.

Nora, Simon, and Alain Minc. 1978. *L'informatisation de la société*. Paris: Editions du Seuil.

Norman, Donald. 1988. *The psychology of everyday things*. New York: Basic Books.

O'Connor, Brian. 2005. *Adorno's negative dialectic*. Cambridge, MA: MIT Press.

Ohashi, Ryosuke. 1997. The world as group-theoretical structure. Unpublished manuscript.

Olafson, Frederick. 2007. Heidegger's politics: An interview with Herbert Marcuse. In *The essential Marcuse*, ed. A. Feenberg and W. Leiss, 115–127. Boston: Beacon Press.

Oudshoorn, Nelly, and Trevor Pinch. 2005. *How users matter: The co-construction of users and technology*. Cambridge, MA: MIT Press.

Palmer, Richard. 1969. *Hermeneutics*. Evanston, IL: Northwestern University Press.

Picon, Gaetan. 1956. *Balzac par Lui-meme*. Paris: Seuil.

Pigeat, Henri. 1979. *Du téléphone a la télématique*. Paris: Commissariat General Au Plan.

Pinch, Trevor, and Wiebe Bijker. 1989. The social construction of facts and artifacts: Or how the sociology of science and the sociology of technology might benefit each other. In *The social construction of technological systems*, ed. Wiebe Bijker, Thomas Hughes, and Trevor Pinch, 17–50. Cambridge, MA: MIT Press.

Pinch, Trevor, Thomas Hughes, and Wiebe Bijker. 1989. *The social construction of technological systems*. Cambridge, MA: MIT Press.

Plato. 1952. *Gorgias*. New York: Bobbs-Merrill.

Preston, Beth. 1998. Why is a wing like a spoon? A pluralist theory of functions. *Journal of Philosophy* 95 (5): 215–254.

Radder, Hans. 1996. *In and about the world: Philosophical studies of science and technology*. Albany: State University of New York Press.

Ricoeur, Paul. 1979. The model of the text: Meaningful action considered as a text. In *Interpretive social science: A reader*, ed. P. Rabinow and W. Sullivan. Berkeley and Los Angeles: University of California Press.

Rockmore, Tom. 1992. *On Heidegger's Nazism and philosophy*. Berkeley and Los Angeles: University of California Press.

Rosenberg, Nathan. 1970. Economic development and the transfer of technology: Some historical perspectives. *Technology and Culture* 11 (4): 550–575.

Rowe, R. D., and L. G. Chestnut. 1986. *Oxidants and asthmatics in Los Angeles: A benefits analysis executive summary.* Prepared by Energy and Resource Consultants, Inc. Report to the U.S. EPA, Office of Policy Analysis. EPA-230–09–86–018. Washington, DC.

de Saint-Just, Louis-Antoine. 1968. *Oeuvres choisies.* Paris: Gallimard.

Schivelbusch, Wolfgang. 1988. *Disenchanted light.* Trans. A. Davies. Berkeley and Los Angeles: University of California Press.

Schürmann, Reiner. 1990. *Heidegger on being and acting: From principles to anarchy.* Bloomington: Indiana University Press.

Searle, John. 1995. *The construction of social reality.* Cambridge, MA: MIT Press.

Seidensticker, Edward. 1983. *Low city, high city.* New York: Knopf.

Shaiken, Harley. 1984. *Work transformed.* Lexington, MA: D. C. Heath.

Sharp, Lauriston. 1952. Steel axes for Stone Age Australians. In *Human problems in technological change,* ed. E. Spicer, 69–90. New York: Russell Sage.

Simondon, Gilbert. 1958. *Du mode d'existence des objets techniques.* Paris: Aubier.

Simpson, Lorenzo. 1995. *Technology, time, and the conversations of modernity.* New York: Routledge.

Spinosa, Charles, Fernando Flores, and Hubert Dreyfus. 1997. *Disclosing new worlds: Entrepreneurship, democratic action, and the cultivation of solidarity.* Cambridge, MA: MIT Press.

Stone, Allucquère Rosanne. 1995. *The war of desire and technology at the close of the mechanical age.* Cambridge, MA: MIT Press.

Suchman, Lucy. 2007. *Human-machine reconfigurations.* Cambridge, UK: Cambridge University Press.

Tenner, Edward. 1996. *Why things bite back: Technology and the revenge of unintended consequences.* New York: Alfred A. Knopf.

Thompson, Paul. 2006. Commodification and secondary rationalization. In *Democratizing technology: Andrew Feenberg's critical theory of technology,* ed. T. Veak, 112–135. Albany: State University of New York Press.

Todorov, Tzvetan. 2007. Avant-gardes & totalitarianism. Trans. A. Goldhammer. *Daedalus* 136 (1): 51–66.

Traweek, Sharon. 1988. *Beamtimes and lifetimes: The world of high energy physicists.* Cambridge, MA: Harvard University. Press.

Tremblay, Jacynthe. 2000. *Nishida Kitaro: Le jeu de l'individuel et de l'universel*. Paris: CNRS Editions.

Turkle, Sherry. 1995. *Life on the screen: Identity in the age of the Internet*. New York: Simon & Schuster.

United Kingdom. 1844. *Hansard parliamentary debates*, 3d ser., 75 (February 22–April 22).

U.S. Environmental Protection Agency. 1999. The benefits and costs of the Clean Air Act, 1990 to 2010. Appendix H: Valuation of human health and welfare effects of criteria pollutants. http://yosemite.epa.gov/ee/epa/eermfile.nsf/vwAN/EE-0295A-13.pdf/$File/EE-0295A-13.pdf.

Venkatachalam, L. 2004. The contingent valuation method: A review. *Environmental Impact Assessment Review* 24 (1): 89–124.

Verbeek, Peter-Paul. 2005. *What things do: Philosophical reflections on technology, agency, and design*. University Park, PA: Penn State University Press.

Wajcman, Judy. 2004. *Technofeminism*. Cambridge: Polity.

Weber, Max. 1958. *The Protestant ethic and the spirit of capitalism*. Trans. T. Parsons, 181–182. New York: Scribner's.

Weckerlé, Christian. 1987. *Du téléphone au Minitel: Acteurs et facteurs locaux dans la constitution des images et usages sociaux de la télématique* 2 vols. Paris: Groupe de Recherche et d'Analyse du Social et de la Sociabilite.

Whitehead, Alfred North. 2004. *The concept of nature*. Amherst, NY: Prometheus Books.

Willinsky, John. 2006. *The access principle: The case for open access to research and scholarship*. Cambridge, MA: MIT Press.

Winner, Langdon. 1986. *The whale and the reactor: A search for limits in an age of high technology*. Chicago: University of Chicago Press.

Winner, Langdon. 1991. Upon opening the black box and finding it empty: Social constructivism and the philosophy of technology. *Science, Technology & Human Values* 18 (3): 362–378.

Winograd, Terry, and Fernando Flores. 1987. *Understanding computers and cognition*. Reading, MA: Addison-Wesley.

Woolgar, Steven. 1991. Configuring the User: The Case of Usability Trials. In *A Sociology of Monsters*, ed. J. Law, 58–99. New York: Routledge.

Wynne, Brian. 1989. Sheep farming after Chernobyl: A case study in communicating scientific information. *Environment*. 31 (2) 10–15, 33–39.

Index

Printed in the United States
by Baker & Taylor Publisher Services